CROSSING THE SCIOTO
ORPHAN'S JOURNEY

Gerald R. Wheeler

Table Of Contents

Dedication

I dedicate this book to my wife Carol Wheeler, Shelley Wheeler Moore and children

Acknowledgments

I would like to acknowledge my mentors and teachers

About The Author

Gerald R. Wheeler Born Jan. 3, 1940, in Columbus, Ohio, holds a B.A. from Long Beach State College, M.S.W. from The Ohio State University, and a Ph.D. from the School of Social Service Administration at the University of Chicago. He served in the Medical Service Corps in the Ohio National Guard and the U.S. Army Reserve, co-founded the Long Beach Free Clinic in the 60s His career included community organization post the Watts riot, activism in juvenile justice in California and Ohio. He held positions as Director of Research at the University of Maryland's Institute of Criminal Justice and Criminology, and Harris County Community Supervision and Corrections Department, Directed Harris County Pretrial Services. His research appeared in over seventy scholarly journals, and books including COUNTER DETERRENCE (Nelson-Hall, 1978). NORTH AMERICAN REVIEW, LOUISIANA LITERATURE, AETHLON, ONTHEBUS, KALEIDOSCOPE His award-winning fiction, poetry, and equine photography have appeared in CONCHO RIVER REVIEW, HORSE OF KINGS, EQUINE JOURNAL DRESSAGE, COWBOYS & INDIANS MAGAZINE. His work has been nominated for the Pushcart prize and the Spur Award for poetry from the Western Writers of America.

1940

Born In The Carney Clan

Coexisting with the shadow

Of the ghost of an uncle,

Never met him, yet

I could not escape other's memories of him

Destined to live out his dreams,

dreams choked by a life cut short

Uncle Mutt held me in a hammerlock whenever I ignored farm chores, rubbed his knuckles into my scalp, calling me, "Knucklehead, Knucklehead" The rubbing continued, pressure mounting until I hollered—unable to take it anymore, "Uncle.". He finished the lesson by wrestling me to the ground with a silly grin on his face. After a few of his lessons, I vowed never to ignore chores again.

However, in public or when he was pleased with me, he called me "Master Gerald" as if I was the offspring of royalty. Not an orphan child, sired by a man whose name was taboo, perhaps partially because my mom, whom I had yet to meet, went insane while married to him. Whether his influence led to her insanity or whether she had a little-understood mental illness, didn't matter. He was to blame in my family's minds. Perhaps the mix of Uncle Mutt's high expectations with reverence toward me was because I reminded Uncle Mutt of his oldest brother Jim whom he looked up to.

At the end of every day, Grandma Carney met me at the door and squeezed the breath out of me like she was grateful I had made it home. But the day she found coal dust on my jeans and boots; I was grounded for a week. For quite some time, she would not take her eyes off me. Uncle Mutt later told me why. I violated Grandma Carney's steadfast rule: ***The mine area is out of bounds.*** A condemned landscape of rocky terrain, hills, open pits, tunnels, and shafts strewn with rusted ore carts and abandoned mining equipment that succumbed to economic hard times and corporate mining competition, beckoned adventurous kids like me. It was full of relics of the *Carney Coal Company* that prospered during *The Great War* era. But truth be told, our family's enterprise was doomed by the traumatic death of Grandmother Carney's first-born son Jim crushed in a mine cave-in at age 19. His older brothers carried his body in a wheelbarrow down the hill to the barn so grandma couldn't see his broken body until it was prepared by the undertaker. According to the family, Jim was smart as a whip, handsome as Valentino, and grandma's favorite child. His death sent a shock wave through the family. It broke grandma's heart. As the wave swelled, it scattered most of the Carney brood to less dangerous endeavors.

Grandma often gazed at and called me, "Jim," before shaking her head, saying, "Ahem, Jerry..." Then, as if in a trance, she would open her leather change purse, worn like a uniform from morning to night, and retrieve Uncle Jim's U.S. Army Rifle Shooting Medals, caressing them like one might their most precious child.

I never corrected her.

Grandma Carney was blessed with nine children who made it to adulthood. Only one, Ralph the runty youngest of nine siblings still lived at home. With five strong older sisters and the runt of four brothers, Uncle Ralph was given the moniker, "Mutt." He never finished third grade and rarely left Grandma Carney's side but loved to tell the same stories over and over again.

Mutt wanted to join the army. He failed the army's physical because he slightly limped from a bad hernia and had a missing testicle he called "seed." His face was the color of leather and creased beyond his young adult years. Eyes perpetually squinting at people, Mutt chain-smoked *Camels* which were secured in his bib overalls pocket. If an ashtray was not within reach inside a room, he'd tap the dangling ashes in a rolled-up pant cuff, even when Grandma Carney was nowhere to be found. While he had severe learning disabilities– being barely able to read and stammering when he spoke– he was an excellent farmer and rancher. His brother Walter was a technical genius; Joe, another brother, possessed a sharp wit of a successful salesman; Mutt never shunned duty on the homestead during economic hard times, family deaths, World War II, and thereafter.

Mutt taught me about managing livestock: harnessing a mare plow horse, raising calves, caring for the chickens in a coop, and making feed store supply runs in a horse-driven wagon. As a jack-of-all-trades, Mutt also passed along his knowledge of planting vegetables in a row, carving a slingshot, tracking a fox, shooting a rifle, fishing, skinning rabbits, and making hides.

One warm spring morning, he brought me with him to plant spinach. I clutched the bag of seeds in my hand, wanting to prove to Uncle Mutt that I could be useful. He carried the hoe over his shoulder, resting in the crook of his neck. He opened the gate to the cow pasture, let me in, and carefully closed it behind him. The dewey grass quickly soaked our shoes as we walked across to the garden.

"Master Gerald," Mutt said, stuttering. "I a got some advice for you."

"What's that?" I asked. I looked up at him, waiting for the next piece I had heard hundreds of times, but ready to play his game.

"Never step twice in the same pile of shit."

Unbeknownst to Grandma, I accompanied Mutt in the surrounding hills to blow up rocks with dynamite. Had she known, both of us would be on the receiving end of her wrath. Dynamite mining yielded coal used for fueling our kitchen stove.

"Now Master Gerald, see here, you a gots to put holes here." He said as he drilled holes in the cave wall, striking steel drill with a hand-held hammer. Carefully, he connected a long fuse to a dynamite stick the size of a flare, inserting the explosive in the hole.

"How long we got?" I asked, pointing to the fuse he uncoiled, placing the line in the path of rubble and stone as we retreated from whence we came.

"Not long." He whispered, "Master Gerald, time to hunker down," as if the volume of his voice would ricochet off hidden tunnel walls and detonate the deadly powder. "Get a back with Red."

I returned to our steam shovel and hunkered down with Red, our mare. Luckily, a blast couldn't spook her because she was used to coal miners' work. Otherwise, Uncle Mutt wouldn't have let my six-year-old hands hold her harness.

"Fire in the hole!" Mutt hollered. Then, a loud explosion rocked the hillside. After the smoke cleared, I led Red to the entrance of the tunnel where I found Uncle Mutt swinging his pickaxe into the fallen rocks, revealing chunks of coal. I ran around gathering the dusty black chunks and filling Red's saddlebags until they couldn't fit a pebble.

"Now, Master Gerald, you did good work on our secret mission," he said and clapped me on the back. "Yep," I said. I winked at him.

"What we goin' to tell Ma Carney?" he asked.

"I was playing in the river."

"Good boy," he said. We both knew the mine area was strictly forbidden for the Carney's youngest and most precious grandson.

My grand folks had no car; cars were mainly owned by middle and upper-class families with money to spare, and they could barely cover their expenses. Uncle Walter, the oldest Carney son, established a prosperous air condition service in Perry and Fairfield Counties and was the first to purchase one. In the 1940s, Walter picked up a trade learned in correspondence courses which blossomed into several patented inventions that preserved food for the war effort, exempting him from military service. Walter married a frail, local woman named Blanch who was an overly protective mother of two children, Mona, and David. They lived in a modern house in the middle of town. Not only did they own a car, but they also had the latest technology: a player piano that played music on a metallic roll, and two telephones including a fancy wall phone next to the upstairs commode where Uncle Walter "conducted his business."

Their son David, my age, was a momma's boy kept away from the Carney homestead. I didn't understand my cousin and his adherence to rules. At age five, I had a few rules governing my daily activities. Rolling hills, woods and streams, and secret caves were there for me to explore. Most days, the wind scooped coal dust from the hills, flowed down the mountain like scree, and saturated our air, mixing with the scent of manure from the barn and the freshly turned earth from the pastures awaiting

Uncle Mutt's green thumb. I loved interacting with the paradise that was my world while David was kept inside, learning to be a gentleman. Even though David couldn't visit us, we benefited from Uncle Walter's largess; we were thankful for a party-line telephone, abundant firewood, and coal to fuel the kitchen stove and fireplace, running water, a working pump well, two flush toilets, and arable land to grow our food and livestock feed.

This was my family. I did not know my father's full name for years. Mentioning him was taboo. Accidentally coming across a photo with him hidden in the back between my mother and her many siblings could not happen; no photo containing his image was kept. My natural father was ostracized by the Carneys—why, I didn't know, and no one filled in the blanks. I had no conscious memory of my biological mother; I only knew that her name was Kathleen, and she was hospitalized because of "*bad nerves.*" My surname of Wheeler did not appear to matter because my dark hair, brown eyes, pug nose, wide Appalachian face, and Irish temper had the "Carney stamp."

I never felt the stigma over being an orphan, or experienced feelings of parental rejection. Though looking back through foggy memories of being a young boy, I recall being led by the hand or carried by kin into cars to sleepovers with cousins to give "grandma a rest." A few days later, I'd be dropped back off at the Carney homestead, shadowed by Uncle Mutt, carrying out my regular chores: tending a vegetable garden, shucking corn, gathering eggs, feeding chickens, and an Angus calf I named Blackie. I filled water troughs, cleaned the barn, and groomed Red, our mare after plowing.

As with many Appalachian boys, I learned to shoot a shotgun before I could read. One day, Uncle Mutt decided I was ready for this rite of passage. Wordlessly, he stepped behind me, and raised the heavy shotgun, placing its butt snugly against my right shoulder, my left hand on the forearm. Gritting my teeth, I cocked both hammers, pressed my thumb against my cheek, and aimed both barrels at a milk can afloat in a murky pond. Trying to control my trembling body, I focused intensely on the task in front of me. *I can do it,* I told myself. With my eyes locked on the target, the weight of the shotgun buckling my knees, I braced myself against Mutt's body for strength.

Mutt said, "Hold your breath, Master Gerald. Starting with the front trigger, squeeze... both triggers."

I held my breath and squeezed the triggers as he instructed.

The dual discharge's deafening blast and recoil lifted the shotgun, pointing it skyward. The blast left my ears ringing, then forced me against Mutt's unyielding form. He took over, lowering the barrels. And with his assistance, I broke the shotgun open, simultaneously ejecting spent warm shells in my hands. Savoring the moment, I held each empty casing against my nostrils, and inhaled the scent of cordite, completing the firearm stage.

When it came to my safety and welfare, Grandma Carney doted over me. Before I left the house to start morning chores, she'd call me into the kitchen for an inspection. Dressed like a pioneer woman transported straight from the 1800s, Grandma Carney always wore her gray hair in a bun that danced as she kneaded bread or molded pie crusts, a long prairie dress protected by an apron covered in flour. I walked into the kitchen, stood behind her, and waited for her approval. When she could pause her baking, she straightened her stooped shoulders, motioned me to move closer, wiped her hands on her apron, adjusted her wire spectacles, then stared at me.

"Hmm," she said, trailing off and focusing her blue-green eyes on the area where my cowlick that refused to be licked into place was. Before I bolted out the door, she eyed and fastened loose buttons, made sure overall cuffs were even, and shoelaces double-tied.

"You'll do."

And with that, I was off for the day, only leaving the patio door swinging in my wake.

Lucky for me I was feeling well. Her eagle eyes could tell when I had even the hint of a cold or an injury. When I was ill, she administered home remedies like applying special salve to the chest to ward off bad omens and cough, doctored boils with hot pads until the pus disappeared, and extracted baby teeth with a yank of string. She sliced holes in leather shoes to relieve the pain of bunions and corns bulging feet and toes. She never scolded me for refusing to eat slimy egg yolks, mush, or lima beans, pilfering cookie jars, drinking milk from the bottle, sleeping in my socks, pounding my head senseless on the pillow until asleep, leaving toys or dirty clothes scattered in the bedroom. She never carried out the threat to "wash out my mouth with soap for using the Lord's name in vain." However, at day's end, she was unyielding in my attempts to skip a *"Duzz and Babbo,"* and leave a bathtub ring, and always checked for potatoes growing in my ears at the bathroom door. She never complained or gossiped, instead allowing me to be the inquisitive, wild child I was.

The quietest member of our makeshift family was Gramps Carney. Gramps was six foot plus, had silver hair, and deep brown eyes like mine. He wore *Dickie* trousers, work shirts, and black high-tops. He rarely ventured outdoors because of a bad heart and painful arthritis from mining his family's strip coal mine until it closed. His ailing body didn't allow him to explore with me, but he could still carve toy trucks with moving wheels and beds that dump dirt and stones. He fashioned sailboats with rudders, keels, and masts with sheet-patch sails that tacked against the knee-deep creek after a storm. We'd sit on porch steps, shaving kernels off shucked corn with razor-sharp kitchen knives to accommodate grandma's false teeth, while he whittled and talked. Or we sat in the sunroom, him in his favorite chair--a Boston rocker, me on his knees for a game of "find the key." I'd close my eyes, inhaling the aroma of apple or peach pie coming from the kitchen as he hid the key to the tool shed Master lock. I'd open my eyes, peek into his shirt pocket, under a cuff, a slipper, pry iron fingers away from his strong callus hand. Sometimes, he slipped the key into my bib pocket or boot and tickled me. Gramps knew how to lull me to sleep. Once I couldn't keep my eyes open, Uncle Mutt carried me upstairs to bed and I'd hear grandpa's violin in my dreams

"Master Gerald, let's go to town."

"Yea," I practically screamed.

My favorite excursion was riding the manure wagon hitched to Red, driven by Mutt to Somerset, the local town. The town was the county seat of Perry County, and located a mile up the road, but it felt farther. I could hardly wait. Once we stopped in front of the country store, I'd scramble off the bouncing seat, racing in the opposite direction, and find myself in the center of a traffic circle staring at a life-size bronze memorial of the town's Civil War hero--General Phil Sheraton. The sign below proclaimed his heroic contributions. He led his fellow Yankees to charge against retreating Rebels. I stared at him while Imaging mini balls in motion whizzing past old glory's stars and stripes flapping in the looming sky, I'd touch the sculptured cavalry boot spurs in stirrups, feel the holstered pistol, gaze at the steel body erect in the saddle of rearing steed, his arm waving a flat top hat celebrating Union victory.

"You're born into a family of heroes," Gramps told me one day he caught me staring at the General.

"Are we related to him?" I asked, hoping to be related to something so grand.

"No, someone even better," he said. He boasted of the Carney Clan's importance during the war, telling truly fantastical tales of our involvement. Most centered around the exploits of Michael Carney, who enrolled on the 2nd day of March 1862 in Co. G-61-Regiment of Ohio Volunteers. Due to his impressive actions, he survived the Battles of Gettysburg, Lookout Mountain, and the general carnage of the Civil War.

After those stories, whenever I looked at the General, I saw Michael Carney: war hero and my relative. I wanted to be as great as him.

1945-1950

Logansport Indiana

Mom, Mother, Kathleen,

Living with the impossibility of

a three-sided coin

Yesterday heads

Today tails

Tomorrow, a mystery

A river flows fast when

It comes down a mountain

With little fanfare, I was sent to Logansport, Indiana where I met my biological mother, Kathleen, and stepfather I called "Earl." As I stood in the kitchen meeting them for the first time, I couldn't help but stare at my mother. *Never*, I thought to myself, *could I ever call this strange woman mom*.

Kathleen leaned weakly against the sink, her rust-colored and matted hair falling over her face, while self-consciously closing her house robe and occasionally flashing her pale nakedness underneath. Like a suddenly caged bird searching for an escape, she flitted her eyes from the strange boy standing before her to something off in the distance apparition. Yet, I didn't even know how to solve the mystery of my sudden appearance or her affliction, both bars that held her imprisoned. Her husband Earl did most of the talking. He was a railroad conductor. When he went to work punching tickets on the train, he'd shuffle me to his folks down the road.

Earl was never abusive toward me or impatient with Kathleen. He taught me to throw a baseball, catch and bat left-handed because he was a southpaw, and how to tell time on his official railroad pocket watch, "the most accurate timepiece in the world." He enjoyed taking me through cabooses, and steam engine locomotives, and meeting railroad workers like brakemen, firemen, and yardmasters. And, he emphasized, above all else: "Respect other people's property."

But he was gone more than he was home. During one of his long absences, I found myself in front of the local *Five & Dime* store. I cruised the hallways, fingering the penny candy, touching the hunting knives, and wishing I had money to purchase a few treasures. When I walked down the gun aisle, I barely thought as I tucked a cap gun and ammo under my shirt. With my heart beating hard, threatening to leap out of my chest, I walked out of the store. I expected the store owner to run after me, but no one followed me. Breathing a huge sigh of relief, I ran down the road to my new home. I set out my new treasures behind the garage, creating a target practice. On a large log, I piled small sticks, rocks, and cans. This is where Earl found me later that afternoon.

"I did it. I stole it," I confessed to him when he saw the loot. I was scared he would take a belt to me bare-legged or hit me with a broomstick like other boys I heard got whooped by dads for stinky fingers or backtalk. To my deep relief, he marched me into the store and made me return my bounty and apologize to the clerk. And that was that.

On another occasion when he was away punching tickets, I started exploring his fishing gear. In his tackle box, he had a shiny but razor-sharp flay fishing knife. After examining the sides of the knife, I got a great idea. I examined downed wood chunks until I found the perfect piece then used the knife to whittle a little boat. Not being much of a whittler, the knife slipped and flayed me instead of the fish it was meant for. I nearly cut off my left index finger. A neighbor saw me leaving the garage, holding my bloody hand wrapped in the bottom of my T-shirt, and rushed me to a hospital. That night at dinner, I hid my mummy-wrapped hand under the table, ashamed and unable to manipulate utensils. Earl silently reached across the table and cut my meat, as my mom sat next to him, staring at her food, fidgeting as usual in her seat. Yet, he didn't say a word about my hand or the hospital visit.

For as much as I liked Earl, I deeply missed the Carney homestead: The smell of coal ore, freshly turned earth, the sight of farm animals, scavenging crows, roosters sounding alarms, and nature's creatures battling extinction. Instead, I sat on the porch alone playing marbles by myself, chewing bubblegum, and playing with collect baseball cards and comic books. One particularly lonely day, my Aunt Helen, Grandma Carney's oldest daughter made a surprise home visit. She looked the exact same as I remembered during her past welfare checks at New Lexington; she had rosy cheeks, coiffured dark hair, wore a flowery dress, and nice shoes, and brought me hugs, kisses, and gifts. She had married an important doctor in Columbus and brought expensive gifts: a blackboard, an easel, colored chalk, and a jar of change. She paid me a nickel for each letter of the alphabet I could recite correctly and write legibly. After a few trials and successes, I graduated to numbers and was awarded in dimes and quarters which I promptly shoved in my pockets. I shared my collection of baseball card heroes; Aunt Helen ooh-ed and ahh-ed in just the right places. When I offered her a piece from my prize pink bubble gum stash, she didn't hesitate. Together we chomped and blew balloons bigger than our faces, exploding on our cheeks. Our chalk-dusted fingers peeling away at the sticky mess left colorful streaks of new skin.

Though she was shorter than Mutt and more refined, she reminded me of home. She did bring the sad news of Grandpa Carney's death. It was peaceful, she told me. Didn't feel a thing. My six-year-old brain couldn't comprehend the finality of death. But then she hugged me close and told me she had to go.

I sadly watched her slide behind the steering wheel of a shiny black *Ford*, crane her neck to see me waving goodbye, feeling a pit of loneliness form deep in my stomach.

Kathleen, my primary caretaker, was anything but caring. I took care of her more than she took care of me. Most days my mom hibernated in the bedroom, curled in dirty sheets, talking gibberish. I preferred this side of my mom. At least, I didn't have to worry about her unpredictability. Sometimes, her eyes would blink fear and I'd know that something awful would happen. She ran upstairs to do her hair, make it stand on end like an Ozark hillbilly woman, apply her lipstick, smear it across her face like war paint. When she returned downstairs wearing only a half-slip and brassiere, she started running around the house, barefoot, to escape her prison.

"Kathleen, stop," I breathlessly said. She didn't gift me a brief glance, just continued to search for a way out at every open window or door. I raced around her, closing curtains, locking doors. Eventually she gave up, returning to her bed. I sat on the end of her bed, catching my breath while she softly snored behind me.

A few days later, I went looking for her because I hadn't seen her in hours. I found her drowning in a bubble bath. Averting my eyes, I grasped her soapy arms to balance her rising, wobbling legs slipping on sleek porcelain. I wrapped her shivering body in towels.

She mumbled, "My nerves," her teary eyes cast downward.

This was typical Kathleen. If she interacted with me at all, it was to make excuses for burning my St. Bridget's school shirt with iron or to apologize for feeding us nothing but crackers, macaroni, and cheese until Earl returned from a railroad trip. Most of the time she floated around, completely oblivious to my presence, never acknowledging who emptied the overflowing ashtrays from her chain-smoking, noticing when I quickly stubbed out her discarded cigarettes she flung down while cooking.

Earl and I never spoke of these episodes, though they became more frequent the longer I lived with Kathleen. Sometimes Earl's family would come to take care of her, so I would get a brief respite from the turmoil, or the nuns at my school called authorities to take poor Kathleen away.

I stopped believing in God before I could recite a tenet in catechism, count rosary beads, write curse words, and ask questions I knew were without answers. I renounced the Trinity and Blessed Virgin the rainy morning a priest whisked me from St. Bridget's the first week of second-grade class. A compassionate grade promotion for school nap time storytelling my homestead adventures with Uncle Mutt no one believed. in. I followed him out into the hallway. After he shut the door firmly behind us, he abruptly turned around and fixed his eyes on me.

The priest said, "Gerald, I am sorry your mother can no longer take care of you."

I nodded.

"I'll make sure you will get to where you need to go," he said. I grabbed my things and left with him.

I thought he was taking me again to my stepdad Earl's backup family in town. But the priest drove directly to Logansport train station and pinned a note on my shirt pocket- *1145 Mulford Rd.*

Grandview Heights-my Grandma Carney's address. He handed me a train ticket like it was another failing report card and a wad of dollar bills.

He said, "Use this to pay for food on the train and taxi ride from Columbus Union Station to your grandmother's."

I could hardly believe my luck. My hopes were coming true. It didn't matter that I was leaving Logansport, Indiana like an outlaw who had been run out of town, with only my clothes on my back. A tiny part of me wished for the option to retrieve Grandpa Carney's handcrafted toys, lead soldier collection, cap guns and holster, pocketknife, and baseball cards, but it was okay. I was going home.

Troop Train

Bombs and bullets explode,

only in my head

Soldiers taking the train home,

Safe at last,

rolling the dice,

tossing coins, matchsticks, and dollar bills

on blankets spread down the aisle

A man dressed like Earl, boosted me up a stepstool leading to the train car. Suddenly, I was climbing over bulging olive drab duffle bags in a train car packed with GIs returning from Victory in Europe. Soldiers wearing Army dress greens and khaki uniforms with shiny brass buttons, colorful patches, insignias, and battle ribbons sat in every seat. These young men openly celebrated by drinking beer and passing flasks of whiskey. Many small groups huddled over mini-gambling operations. Some rolled dice, while others carefully fingered hands full of Jokers, Aces, Queens, and Kings before tossing dollar bills in the aisle.

I found the last empty seat. The car trembled and shook, then slowly picked up speed down the tracks. I choked on cigarette smoke. Bombs and bullets exploded in my head. Soon the rhythmically clanging wheels and shimming of the train cars put me to sleep just like long car rides to visit kin do.

When I awoke, a thin, blond-haired soldier was standing in the aisle by my seat. He was removing an Eisenhower jacket that had three gold chevrons on each sleeve and a green shoulder patch with the bold blood-colored numeral, **1**. Each collar lapel had a shiny brass crossed-muskets Infantry insignia. The soldier carefully folded and secured the jacket in the overhead storage area.

"Hey, kid, where you going?" the soldier said.

"Co..., Columbus."

"Aren't you a little small to be traveling by yourself?"

"I'm six, almost seven."

"What's your name?"

"Jerry."

He reached out and shook my hand firmly. "Jake. I'm going to Cleveland." He had dark bags under his deep-green microscopic eyes.

"You fight in the war?" I asked, recalling neighbors firing guns in the streets celebrating victory on VE Day. He nodded, looking back at the car entrance as if my folks were late arriving passengers.

I was too embarrassed to admit my dad was missing in action. A deserter so low or bad the Carney clan denied his existence. Now facing real war heroes returning from battle, I created a fantasy father to deflect inquiry and conceal my secret shame. I lied.

"My dad was a *B-17* bomber pilot killed by Nazis."

The soldier's eyes softened. He gently touched my shoulder, "I am a sorry kid," he said, reaching down to the duffle bag propped between us.

Heart pounding, my eyes locked on his callous hands opening the olive drab canvas bag, retrieving a scarlet flag with white circles containing the infamous black Nazi swastika symbol.

"I got this off a Tiger tank," he said, tying it around my neck like a scarf. Then he produced a silver German infantry belt buckle and Iron Cross, placing them in my hand for inspection. Next in his Mary Poppins bag, he pulled out a stiletto with a silver snake and the letters *SS* on the hilt and a German Luger wrapped in oil rag. He ejected an empty clip and let me hold it. The infamous pistol felt heavy and as valuable as gold. Lastly, he shared a photograph of himself and GI buddies standing in front of a Nazi tank that had hand-painted bold words, ***Kilroy Was Here*** on the turret

"This here," he said as he pointed to the words, "is the GI's calling card as they advance in enemy territory." He was a true hero. Not my dad.

I disembarked at Columbus Union Station in the pouring rain, a GI helmet liner tilted over my ears, a swastika flag scarf knotted around the neck, a cartridge belt with pouches filled with candy and chewing gum slung over slouched shoulders like a child soldier. Jake stood up to talk with some of the other soldiers on the train after he showed me his gear. I stared out the window, wishing my dad was like Jake.

Soon, he was back, friends in tow. Before I could correct my lie, they lavished me with souvenirs, each one coming with a story. They were drunk, but they clearly felt sorry for a kid whose father died fighting Nazis. A porter directed me to a taxi.

I stepped into the taxi and shut the door. The driver reacted like I was a lost kid until I handed him the note and wad of bills. I fell asleep listening to the rhythmic sounds of rain pelting metal shells yawing in a flooded street, windshield wiper blades repelling waves from a black sky. Unwillingly pulled from a deep sleep, I felt something shaking me.

"You're home, kid," a loud voice said, muffling machine gun bullets hitting the car's roof.

My eyes squinted through the window at sheets of rain reflecting in beams of the taxi's searchlight illuminating numbers, ***1145*** above a small porch. Home I was. I had not seen Grandma Carney for over a year since I left the farm and started school. *Now that Grandpa Carney was gone,* I thought to myself, *she needed me*. I exited the cab, feet splashing flooded sidewalk, rain pelting my plastic helmet, beads dripping down my back and chest. I dragged myself up the porch steps and pounded on the door blinded by the searchlight beam aimed by the taxi driver. The porch exploded with more light.

Uncle Mutt opened the door with one hand as he snapped a loose strap of his paint-splattered bib overalls with the other. As usual, his body was listed on the hernia side.

"Master Gerald," the familiar husky voice said.

"Hi, Uncle Mutt," I said, standing at attention dressed for war and soaked to the skin.

But that did not matter. He pulled me into the house. Behind him, floating down the stairs was Grandma Carney wearing grandpa's red-flannel robe. When she reached the bottom, she paused, straightened her hairnet, adjusted her wire-rim spectacles, and moved faster than her old bones should have been able to move. Her round face burst into a warm smile. When her strong country hands grabbed my shoulders and pressed me against her soft bosom, she squeezed all my feelings of abandonment out of me. Within minutes, Uncle Mutt removed my helmet, put my head in a hammerlock, and gave me the standard knucklehead rub treatment. The homecoming felt familiar, yet the room felt strange like something was off. I started toward my old bedroom, but Uncle Mutt steered me in a different direction. He led me upstairs as he explained we'd share a room until his younger sister Wanda, asleep in the other bedroom, and destined to join poor Kathleen at Columbus State Hospital by the end of the week, left and freed up a bedroom for me.

The next morning, Aunt Helen arrived in her shiny black *Ford* and her equally shiny exterior. Her commanding presence was made even larger in her tall heels, smartly-dressed, perfect auburn hair curled over sparkling earrings, meticulously applied makeup on the healthy complexion of skin *blessed by Carney genes*, she could have been featured on the cover of *Woman's Day* magazine. She smiled, dropped a bag of new clothes on the floor, and kissed me on the cheek.

Anne's "Mixed-Up Hromosomes"

Coal-black eyes,

staring at me from her humongous head,

wearing unusual clothes:

boy's jeans,

oxford shoes.

Drew attention from the second-grade bully.

He mimicked her until

she punched and kicked him–gave him a licking he'd never forget.

We danced home.

The loss of my toys left behind in Indiana was soon made up by my nine-year-old cousin Anne Carney who lived in an adjoining duplex. She had coal-black hair cut in Dutch-boy style, big brown eyes, jumbo ears, a humongous head, and wore thick glasses tied with a kitchen string. She was sitting cross-legged on her front porch next to a treasure chest overflowing with toys. She was dressed in baggy jeans, a new white jersey emblazoned with the image of Hollywood's famous cowboy hero--*Hopalong Cassidy;* his hands grasping reins of rearing white steed, waving a white Stetson big as his smile. She had a pug nose, boney cheekbones, and a jutting jaw with an off-centered dimple. Her mouth was too small for her buck teeth.

Eyeing me, she pointed at her chest to identify herself, and drooled, "Anne...Anne, ..." Her tongue darted out like a hungry lizard snatching a fly. "Wanna play?" she said.

Pointing at my chest, mimicking back her gesture, I nodded and said, "Jerry."

Anne's big hands retrieved and dumped a drum of *Lincoln Logs* from the treasure chest on the floor and communicated in pantomime to help her construct a cabin. Sitting Indian style, we started erecting a log cabin until she got antsy, stood up, reached down, grabbed my arm, pulled me to my feet, and dragged me off the porch to the backyard swing set.

Wiggling into her seat, she grasped the chains, pumped her lanky legs, and said, "Hoosh,..me. Hoosh...me."

Standing behind her, I took a deep breath, planted my feet, grabbed corners of the seat, pulled her back with all my strength, and heaved. She felt as heavy as a filled feed sack. When she returned from the flight, almost knocking me back to the ground, I planted my feet again and shoved her repeatedly.

Soon, we were both flying high in the sky, parachuting down to earth, shouting, "Geronimo! Geronimo!"

Later that evening, Grandma sat me down to explain that Anne's speech impediments, poor coordination, and unusual physical appearance were caused by her "mixed-up chromosomes," whatever that meant. Anne's Dad, Uncle Joe, was the second oldest Carney son. He was loquacious, jovial, and extremely overweight. He married a Perry County beauty named Evelyn. Evelyn was also his deceased brother Jim's fiancé.

Anne's behaviors and appearance produced stares in public that pained her family. Adults often averted their eyes, and kids made ugly faces or mimicked her gestures. This was so unfair according to my seven-year-old brain. I was an orphan, and no one made fun of me. Except for the incident on the troop train, no one directly raised the issue of why I lived with my grand folks. My orphan status was invisible to strangers unless they followed me home. People like my Uncle Jim die in coal mine cave-ins; adults and children pass away in car accidents; soldiers die or are maimed at war. During rare shopping trips to *Lazarus* department store downtown Columbus with Grandma Carney, I saw legless men on street corners, hunched like stumps on plywood slabs with roller skate wheels for mobility, selling pencils in tin cups. I'd seen magazine and newspaper photos of wounded soldiers and polio victims imprisoned in iron lungs. No one taunted the legless men on the street corners. Their physical conditions were no more self-prescribed than those members of my family haunted with various mental or physical ailments: my cousin Anne's "mixed-up chromosomes," Uncle Mutt's mental slowness, or my mother and Aunt Wanda's mental illnesses. Nonetheless, I found myself clinging to Carney's roots—I was a Carney through and through, mental illnesses and mixed-up chromosomes to be damned. Anne's "mixed up chromosomes" didn't keep her from doing what she loved, and damn my orphan status wouldn't keep me from doing what I wanted to do.

Johnny Fuckerfaster

Kids bully

Kids taunt

Kids tell jokes and make each other laugh

laughter that echoes long,

especially when inappropriate

Grandma signed *Jessie Carney* in the designated *Parent or Guardian* space of the school's papers and that was that. I was going to school.

Robert Lowell Stevenson Elementary School was a short block away from grandma's duplex. Compared to St. Brigit's in Indiana where students were forced to wear uniforms, attend daily mass listening to old priests mumble in Latin, spout boring sermons, and feel the sting of ruler-smacks across the palms for back-talking nuns and horse playing in class, *Robert Lowell Stevenson Elementary School,* was a wild and colorful place. My wide eyes scanned the playground full of girls with pretty ribbons in hair, dressed in fancy jumpers, frilly blouses, pleated shirts, argyle knee socks, gleaming patent leathers and saddle shoes and boys in their smart suspenders and button-up shirts. Despite a record of high absenteeism and failing grades, St. Brigid's gave me a compassionate promotion to second grade. So, I enrolled in second grade at Robert Lowell a month into the term.

I sat low in an assigned back-row desk carved with the initials of dunces on that first day. Hearing my name called, I slowly rose, cringed like a spy captured behind enemy lines, wiped my sweaty hands on prewashed baggy *Levi's*. A lovely, young woman with a face-beaming smile, flowing red hair, and far-reaching eyes–nature's gift- stood at the front of the room looking at me expectantly. I sighed with relief that she was not encased in a starched white habit or a curved body imprisoned like sin in a black robe. She wore a charcoal skirt, matching sweater, and short- heeled shoes.

With a soft voice, she said to the group, "Attention class, today we have a new student, Gerald Wheeler." She paused for a few seconds while glancing about the room before directing her attention back to me, "Gerald, welcome to *Robert Lowell Stevenson."*

I nodded at her, too embarrassed to speak. She must have read something on my face because she quickly turned to the class and started the lesson.

The first time Ms. Zimmerman turned her smile on me and asked me to read a book written in hieroglyphics aloud, I stammered and contracted a terminal case of laryngitis. Undeterred, she plied me with *Smith Brother's* licorice and cherry cough drops and tutored me after school. The time I spent with her was almost as enjoyable as recess time—almost.

Recess provided a completely different type of education. During recess, I would squat in the grass among the trees adjacent to school property, demonstrating my mumblety-peg skills with a pocketknife and laugh at older boys telling, "Johnny Fuckerfaster" jokes that went over my head. One of the older boys would turn to another and say, "Did I tell you about the time Johnny's mother walked into the kitchen, discovered him on the floor on top of a girl, and screamed in shock, 'Little Johnny Fuckerfaster!'"

"What did Johnny say?" a boy would inevitably ask.

"Well, Johnny yelled back, 'I can't, Ma. I'm Humpin' as fast as I can!'"

The group would laugh loudly. I laughed heartily along, but not knowing why.

When Uncle Mutt was not painting houses, which was most of the time, Anne and I borrowed his painting supplies to experiment with colors in the garage. We painted rainbows, mountains, birds, and trees on the interior walls. If the painting supplies were missing, we found other activities to occupy ourselves. On sunny days, we attached strings to empty *Pet* milk cans, flipped them upside-down, and stood on them. Clomping down the driveway, we turned and strutted down the sidewalk like Clydesdales on parade.

By far, one of my favorite memories was when we walked to the nearby confectionary and played pinball and our favorite arcade game—Duck Shooting. Nothing compared to the joy of shooting ducks with a single-shot, compressed-air pistol mounted on a swivel encased in Plexiglas that fired a small pea-sized ball bearing. All this for just a penny per shot. When the "bullet" hit a duck moving across the horizon, the wooden animal tumbled over "dead" into a make-believe pond. After pinning the first mallard, Anne was so excited that she dropped a handful of pennies.

I gave her mine, and told her, "Keep blasting."

She did.

At the end of the week, we looked forward to her dad's Saturday night poker games. Uncle Joe sat hunched over stacks of colored plastic chips, shuffling cards on the Formica kitchen table. To his right sat Uncle Mutt's *Wild Root* slicked-back jet-black hair, gleaming like the sun reflecting on a still pond. Rounding out the group was our disabled ex-marine neighbor and his glass eye, taking the place of the one shot out by a Japanese sniper at Saipan. Empty beer bottles lined the kitchen counter like prisoners of war. Cigarette and cigar smoke drifted like a fog bank from the kitchen into the living room where Anne and I lay prone on pillows, gorging ourselves on popcorn, as we stared mesmerized at TV cowboy heroes: Bob Steel, Lash LA Rue, and Roy Rogers blasting gunslingers and rescuing damsels in distress or the brilliant Charlie Chan solving complex murder crimes. I taught her how to build Lincoln log forts, Finger-paint, and shoot marbles. One day, after her dad gave us five dollars, I took Anne to an amusement park down the street. We threw darts at balloons, rode mustangs, crashed dodgem cars, pelted people with peanut shells from a Ferris wheel, and drowned in cotton candy. After spending the last cent throwing rings for a teddy bear, we tossed our caps, cheered, and danced home.

As fall turned to winter, we found ourselves indoors more. To prepare the house for Christmas, Anne and I fashioned Christmas tree decorations at Grandma's kitchen table. We cut and glued paper rings, strung popcorn with needle and thread, painted walnut shells with watercolors, and rained tinsel on pine branches drooping tubes of liquid bubbling-color lights strung by Uncle Mutt and Grandma. Anne counted down the days to Christmas, starting with "5 days...,4 days....,3 days... to Santa."

Christmas came and went. Soon, I was back at school. On the first day back, I walked in the classroom, then almost walked out again. Instead of a beautiful young redheaded woman writing a lesson on a chalkboard, a skinny, gray-haired lady stood in front of a desk, checking off names on the class roster as students sat in their newly-assigned desks. Before I could ask another student where Ms. Zimmerman was, I heard the gossip whispered around the room.

"Did you hear she was fired for being an unmarried woman with child?"

The loss of attention and much-needed tutoring by Ms. Zimmerman resulted in sinking my interest in school. I did no work. My new teacher sent home a letter informing them of my failures. So alarmed over failing grades, Aunt Helen enrolled me in Grandview's St. Christopher Catholic Elementary School and arranged taxi transportation to and from school. But catechism and three Rs couldn't compete against recess, exploring nearby Grandview Woods, and watching TV westerns and gangster movies with Anne next door. I didn't improve my grades much. Luckily, summer arrived. And with it came no more taxi rides to Mass and boring school.

Grandview Heights was a white working-class suburb of duplexes and small-framed, side-by-side houses sharing space with unattached garages and hiding cars leaking oil in back alleys. Sandwiched between the small houses were ring-size yards with ugly hedges, and random trees uprooting sidewalks and lives. Neighbors included Ohio State Prison guards, factory workers, store clerks, teachers, and secretaries. Most bused to work. A few families like my Uncle Joe's could afford a personal car. Grandma Carney and Uncle Mutt scraped by on her widow's pension, his odd jobs, and family handouts.

But playing outdoors was free; amusement and entertainment was cheap. For two bits, kids sat for hours in darkness at Grandview theater, a mile walk from home, watching double features of pioneers circling wagons attacked by screaming savages, mummies rising from the dead, three musketeers' flashing swords at hapless villains, newsreels of real war and disasters, and *Tom & Jerry,* and *Popeye* cartoons. We would hike through the nearby woods to spend our pocket change at the penny arcade. On rainy days, Anne and I listened to adventures on grandma's radio, mailed cereal box tops to get *Sky King rings*, played tic-tac-toe, sculptured figurines with putty, and finger-painted all colors of the earth.

I still had regular chores to complete, though I preferred my former on the homestead, I finished them nonetheless. I lugged a laundry basket of grandma's scrub board-washed and hand-crank-rinsed clothes up our steep basement steps to the backyard clothesline to wind-dry one day when Aunt Helen made another surprise welfare home visit.

"Honey, pack your jeans. You're coming home with me," she said.

I ran upstairs, chores forgotten, and grabbed my cap-guns, western hat, one-eye teddy bear bleeding straw stuffing from its ears, and a sack of clothes. Dragging my meager possessions behind me, I walked down the stairs and towards the front door, passing the only two black and white photographs we had on the wall: a framed matted regal portrait of Grandma Carney's Mother-Great Grandma Webb, and a tattered picture of boy on a horse. Grandma Webb with her long. braided hair, beautiful skin, piercing eyes wore a white blouse with an ivory brooch pinned on a starched collar, staring off into the distance. The other picture was a shirtless boy in a swayed-back western hat, hands grasping taut reins, a bandanna tied like a bib, legs in borrowed chaps, street shoes jammed in leather stirrups, straddling a tired pony. This second image was taken by a vendor canvasing grandma's neighborhood with a paint pony, chaps, and a box camera attached to a tripod. He charged her $2.00 but did not allow the rider time to retrieve his holstered six-shooters or gallop to his dreams. Sadly, the vendor's equine photography skill was wanting in framing animal subjects. The pony was a black and white paint posing against a dark shrubbery. Part of the pony's head and shoulders were missing and bushy leaves growing out of his ears.

I met Aunt Helen in the front room where she stood with Anne and grandma. Turning to them, I gave them quick hugs goodbye, thinking Aunt Helen was giving grandma a few days of "rest," and I'd be back soon enough.

How could I know at the time that I was leaving our Spartan duplex with windows missing on the alley side, dingy carpet, and faded wallpaper, for good? I was never to remain in any one place for long.

The Surgeon's Den

Some prisons are the prison of poverty,

Some are the prisons of wealth–

both hold you,

tightly

Compared to working-class Grandview Heights, Upper Arlington was larger in every way: wider streets, three- and four-bedroom single-family homes, oversized attached garages, expansive manicured lawns, and "Old Arlington" mansions with winding driveways, sprawling giant trees, and samplings anchored by wire.. It lacked alleys and featured very few sidewalks, The Watsons lived in an ivy-covered, white-painted brick, three-story, custom-built slate roof, residential home adjacent to a vacant lot on *Onandaga Dr* without sidewalks. It had a chimney, fireplace, and attached two-car garage sharing space with garden tools, bags of potting soil, and bikes, with no evidence of spiders, junk, or leaking oil. A flagstone path started at the curb and ended at a crescent-shaped front porch supported by tall colonial columns. One side of the double doors had a mail slot and the other an iron ring-shaped knocker. But the main house access was the kitchen entrance by the garage. Each room had its nomenclature.

The Kitchen was designed for standing room only; all business of cooking: a modern gas stove with open-flame grill, deluxe size refrigerator, metal cupboards and drawers, double porcelain sinks with a window view of the backyard, shrubbery, and vacant lot. With a proud smile, Aunt Helen pulled me into the spacious Breakfast room off the compact Kitchen. Six wrought-iron chairs with deep seat cushions encircled a thick glass-top table. Extended across the wall behind the head of the table were built-in cabinets and drawers for storing dishes and utensils and a counter displaying a fancy radio that had a sliding horizontal dial cover that was closed when not in use. The opposite wall had a bay window view of crab apple trees bordering the driveway and the empty lot. Exiting the Breakfast Room, she paused briefly in a narrow hallway and gestured toward a closed door.

"This is the Green Bathroom," she said, continuing a few paces forward.

I could not say a word.

She motioned us to the side, allowing a quick peek into the Den and the furnished screened-in Sunporch, shaded by green color canvas awnings. We backtracked to the Foyer that had a Mail Table, and a Cloak Room. Off to the other side, we proceeded to the Formal Dining Room, decorated with a large free-standing Japanese screen art panel with painted images of robed Japanese women walking in a floral garden before a mountain horizon. The dining table accommodated at least a dozen guests. It had silver candelabras, sparkling crystal candy dishes, water pitchers and flower vases waiting for fresh flowers, China dishware, and polished silverware in glass-encased cabinets. Executing an about-face, we crossed the Foyer to the Living Room, a misnomer as one observes mostly objects created or once owned by now-dead people: straight-back antique chairs and a writing table used by long-deceased royalty. Even the decorations were from another era: a marble mantle mounted above a fireplace decorated with Louis XIV gilded clock with hands that suspended history, figurines of angels spreading glorious wings, the Virgin Mary embracing the infant Jesus, St. Francis of Assisi feeding imaginary birds, and chubby cherubs looking foolish, listening to Cousin Jane play "chopsticks" on a baby grand piano. My favorite work of art was a four-foot marble sculpture of the naked armless torso of Aphrodite, the Greek goddess of love and beauty, standing recessed on the wall above a fancy oyster color fabric couch. She tempted visitors with her cone-shaped breasts-- nipples, as sharp as tips of a hunting arrow, I later surreptitiously touched.

Scaling the green-carpeted stairs, we continued the tour upstairs. At the top of the landing, we made a quick right turn into her son Jim's bedroom, furnished with a four-poster antique king-size bed, large oak desk, and chair. I had never met Jim, so I stared at the objects in his room, trying to understand him through them. My first surprise was seeing a WWII German steel helmet, a Nazi armband, and sheathed bayonet displayed on a table, and a collection of GI Army Division patches sewn on Olive drab color clothing hung on the wall. But what raised my brows was a looted official **STOP** sign secured on the wall of the immaculate room.

Would we get along? I wondered. *Would we spend all of our time together like Anne and I did?*

We continued, passing the Blue Bathroom, Floral Guestroom , and a clothes chute-- something that I never heard of. She allowed me to peer into her bedroom, complete with twin-beds, enter their daughter Jane's four-poster bedroom, and examine the Pink Bathroom Jane and her parents shared.

Finally, Aunt Helen opened a door across from Jane's room, looked upward, and said with labored breath, "Honey, your room's upstairs. But I can't make the climb because I have a bad foot. Jane will show you your room when she returns from swimming at the club. You'll meet Jim when he returns from summer camp next week."

"Ok," I said.

Glancing at her watch, she said, "Honey, why don't you settle in the den with some magazines while Auntie Helen grills us a hamburger."

Lured by the aroma of pipe tobacco, and a pungent breath of leather, I sank into the surgeon's cracked, blood-red leather chair. I directed my eyes to a cabinet holding pipe racks, a bowl of tobacco, a large round glass ashtray with a cork implanted in the center to knock out pipe ashes, a humidor, and cherry wood wall shelves lined with books.

What kind of books were these? I wondered. I had never seen such large books before.

I pushed myself out of the chair, stood on my toes, retrieved a heavy hand-bound book, and fell back into a deflating cushion, eyes glued to open pages. Image after image of deformed naked bodies-- faces with rotten skin and ugly warts, a man with giant testicles, hunched-back woman with deformed breasts, illustrations of skeletons, muscles flexed and torn—interspersed by names of diseases, medicines, and cures I couldn't read. Though I had skinned rabbits and dissected chipmunks that we shot on the farm and dodged spewing blood of crazed decapitated chickens, these pictures were far more horrific. They made my stomach turn and my eyes look away because the images were eerily human. I replaced the medical book and surveyed shelves lined with Lincoln, Churchill, hand-bound history of war and peace, Sherlock Holmes mysteries, and encyclopedias covering subjects of the world and universe. Along another wall was a window surrounded by glass shelves displaying knick-knacks and golf trophies. The Den included a couch and coffee table containing religious books, a coiled rosary, and a framed photograph of Bishop Fulton J. Sheen in full regalia. Across the room, eyeing the devout reader was an oil painting of Madonna & Child. I sat down in a chair, picking up and biting a curved-stem pipe like the world-famous detective. On the small end table, I found and flipped through pages of *Life* magazine. When I grew bored of that, I stared out at the Den's bay window view of a vast backyard partially shaded by sycamore, maple trees, tall scrubs, and blooming flower beds.

The first meal at my new home was a sumptuous, grilled hamburger smothered in condiments, a slice of red onion and lettuce in a toasted bun served on China *Spode* Camilla with scalloped edges, ridges, shells, tendrils wavy patterns as complicated as my life. The plate was cushioned by a thick fabric placemat next to a white cloth napkin rolled in a silver ring and aligned with silver utensils. I drank milk in translucent sparkling glass, not a faded purple or red aluminum peanut butter jar purchased at Grandview *IGA* that made every drink taste like fucking tin.

"Room For One More"

Coming home dirty from my dugout,

ordered to undress in the basement

"Gerald, bathe for dinner."

My aunt calls.

Dutifully,

I do.

"Jerry's here," Aunt Helen said as my 10-year-old cousin Jane returned from the swimming pool. Jane burst into the house with a towel slung over her shoulders, wet hair reflecting the brightness of the sunshine.

Jane had the contagious Carney smile and friendly eyes. But her hair was muddy-blonde and short, not dark like her mom's and she had a beak nose. She stepped forward, as I arose from my chair, hugged me, then stared at my wild hair as if it were a road-kill toupee. She grabbed my hand and guided me down the hall to the Green Bathroom, turned on the faucet, and splashed water on my locks. Meticulously, she combed my hair until it reflected light in the mirror. Together, we proceeded to the third-floor bedroom that she explained was previously occupied by Checko, a Japanese teen whose father was an important government scientist who, with Dr. Watson's influence, negotiated an alternative to an internment camp for his daughter. The room was painted white, had barren walls, sparsely furnished with a small desk, single bed, chest of drawers, and bookshelves filled with *National Geographic Magazines, Hardy Boys, and Nancy Drew.* The attached bathroom had a shower, commode, sink, medicine cabinet, mirror, towel rack, shelf, and a small window.

Jane pounced on the bed with a wistful smile. "I played with Checko all the time up here. She was like my older sister," she said softly.

"Sorry," I said and walked to the window outside the bedroom. The third-floor window at the top of the stairs offered a panoramic view of vacant fields and a stone-wall fence bordering Lane Avenue and the property of *Scioto Country Club's* golf course. I could see for miles in several directions.

Jane came up behind me and said, "Jerry, you are so lucky. Mother never comes up here."

The rest of the family was not exactly welcoming. Jane's 14-year-old brother Jim was tall, had curly hair, a long nose, a pimply face, quiet, and paid me no mind. He didn't have much interest in playing with his little cousin. Their dad left for the hospital before light and returned usually by six p.m. except Sundays and interacted very little with the family. The first time I met him, it was after Aunt Helen pecked his cheek at the kitchen door.

She said, "This is your Uncle J.O."

He looked down over steel ream glasses as if I was a patient. We exchanged firm handshakes. He had dark hair turning gray, a modest mustache, and wore a suit, bow tie, pale blue oxford cloth button-down shirt, and polished shoes. He appeared physically fit and distinguished as a British prime minister I'd seen in a magazine. The daily ritual rarely varied. After he pecked my aunt on the cheek at the kitchen door, he quietly went upstairs, changed into casual seer suckle trousers, a short-sleeve shirt, and blue *Ked* shoes, then retired to his leather throne in the den, perusing the *Columbus Dispatch* until summoned to dinner. Jim said grace and we ate in utter silence until my cousins asked to be excused. Following the meal, Uncle J.O. returned to the den, smoked a cigar or pipe, read detective novels, and retired by 9:30 p.m.

As the oldest Carney sibling, Aunt Helen was the main beneficiary of the family's strip mine business before the Great Depression. Of the five Carney daughters: Helen, Ruth, Margaret, Kathleen, and Wanda, Helen was the only one educated at the prestigious *St. Aloysius Girls Academy* high school in New Lexington. She married Dr. James O. Watson, a prominent osteopathic physician, and surgeon who successfully fought the *American Medical Association's* (A.M.A) relentless campaign against the legitimization of osteopathic medicine. During WWII, the federal government would not commission osteopaths (DOs) as doctors to serve in the military like M.D.s. Dr. Watson worked with Ohio state politicians, often traveling to Washington to change the law which granted DOs equal status with MDs. He founded *Doctors Osteopathic Hospital i*n Columbus and was the first osteopathic doctor appointed by Republican Governor James Rhodes and subsequent governors to the *Ohio State Medical Board* that licenses all Ohio physicians. His patient list reflected his professional reputation. On weekdays he worked six to six, squeezed in golf mid-morning on Wednesdays, and made hospital rounds on weekends. He looked like Walter Cronkite but never discussed the news, politics, or his work. He rarely conversed with or acknowledged his own children. I called him, "Uncle J.O." I arrived in summer, but entered a mansion entrenched in a permanent winter. Children left their shoes and feelings at the door. They should have answered the phone saying, "Dr. Watson's Residence. Here, we keep personal conversations short."

I wondered if Aunt Helen took me in against the surgeon's wishes. He paid me even less heed than his own children. One night I was late for dinner. My aunt grabbed me by the arm and led me outside the kitchen entrance within earshot of her husband and children eating at the breakfast table. She brandished a switch snapped from a bush. Except for the day Grandma Carney intervened when Uncle Mutt raised his leather belt for "backtalk," I had never been threatened or experienced physical punishment beyond a nun's ruler swatting on palms for horseplay at mass or school. Aunt Helen gritted her teeth and tensed her body, but her soft, intelligent eyes lacked the fire of inflicting physical pain on a child; corporal punishment was against her nature. She lowered and dropped the stick, her attempt to demonstrate to the doctor that she could control a wild child failed. Feeling her embarrassment over losing face as the ruler of all matters regarding her children, I felt ashamed and was never again late for the evening meal.

Besides Jane, the other person I grew close with was Lizzy, the housecleaner. She treated me like a favorite grandchild, always ready with an embrace or a kind word. Never did she disclose the third-floor menagerie of turtles, garter snakes, guppies, or the pungent fumes of paint, thinner and *Testor's* glue used to build model planes, cars, and ships, or ink stains on the desk to her mistress. Every week the bedroom drawers and bathroom shelves in my third floor fortress were magically replenished with immaculate folded clothes and fresh towels.

Country Club Rules

Confusing, unstated rules of wealth

Everyone knows,

but me.

After many mistakes,

I learn–

Navigating the waters

called

childhood

The Scioto Country Club swimming pool was a few miles from the Watson's house via Lane Avenue and the *Scioto River Road.* We entered a rolling landscape of trees, winding fairways bordered by high grass, weeds, and a creek, leading to a manicured patch of shaved grass surrounded by sand traps, staked out by a pennant marking the hole. We saw elevated areas where players practiced swings while caddies washed golf balls and guarded golf bags of clubs. We pulled into a parking space near the clubhouse restaurant, clay tennis courts, a driving range, and a practice putting green.

My aunt twisted around in the car, lowering her sunglasses to peer at us. "Jane, show your cousin around, please," she said.

"Yes, ma," Jane said.

We exited the car, attacked by chlorine fumes announcing the arrival of summer. We saw toddlers and mothers splashing in the baby pool under the supervision of a teenage male lifeguard hunched in a highchair, sucking on a whistle. The regular pool was packed with screaming older kids executing cannonballs off a low board and scaling the ladder of the high board before making a suicidal leap into the water far below. Others ran around the side, engaged in horseplay with towels.

The whole place made me feel like I was going to throw up.

My only previous experience with water consisted of watching Tarzan's swimming in movies. Here I was, about to start the fourth grade and too embarrassed to admit I did not know how to swim. Not only that, I had an aversion to heights. I sat on the deck, dangling my feet and watching Jane and other girls gracefully dive in gleaming blue-green water, stroke freestyle, and attempt a few synchronized movements. They surfaced, gasping for breath, removed their bathing caps, shook the water out of their wet hair, and collapsed in plastic lounge chairs, squinting at the bright sun. They made it look so easy.

Thirsty and dripping water after dog paddling in the shallow end, I wandered over to the refreshment stand feeling a little lonely and sorry for myself. I wish I had money. I wish I knew how to swim. I wish I knew someone other than Jane. Nursing my misery, I started to notice other kids in wet swimsuits, ordering food, and signing a paper slip before retrieving hamburgers, cokes, fries, and ice cream filled paper cups and wooden spoons. Unable to ward off the delicious greasy smells lingering everywhere I went, I succumbed to temptation and committed my first misdemeanor: forging *Dr. J.O. Watson's* name in the blank signature space of *Scioto Country Club Member*.

Walking around the pool, I finally felt like I fit in. Jane made eye contact with me across the pool and swiftly walked over, stopping in front of me. She stood there a moment, eyeing me while I sipped ice-cold Coke.

Jane said, "Mother will be mad."

I grinned at her. "It's amazing," I said after gulping down a mouthful of fizzy sweetness.

She started laughing and quickly became my accomplice.

It took a week to work up the courage to climb the high board ladder, tiptoe to the edge of the slick board, close my eyes, then step into the unknown. Hitting the water and sinking in the deep end, I instinctively kicked hard for the surface while holding my breath. Once my head emerged, I gulped air greedily and madly flailed my legs until I reached the side. Though I was too ashamed to take group lessons with younger members, I wasn't going to let my lack of knowledge keep me from the fun. I never truly learned to swim. But I embraced my new world of wealth and privilege like I hadn't known poverty. Being a summer guest at a private pool with free coke, juicy hamburgers, and ice cream, living in a mansion with hoards of books, glitzy magazines, maid service, three-course dinners at home and regularly going out to fancy restaurants, building model airplanes, evenings listening to *Hit Parade, Sky King, The Green Hornet* and *This is the FBI* with a pretty cousin, was like living in a dreamland. A dreamland topped with perfectly manicured lawns, trees and shrubbery. In the back of my mind, I expected to wake up from this magical dream, returning to Grandma Carney and Uncle Mutt at the start of the school year. After all, they were the only two I had spent more than a year with at a time.

Afraid it would break the spell, I didn't ask my aunt when she would return me home. After all, most previous moves had come without warning. Almost two years ago, a priest suddenly put me on a troop train in Logansport, Indiana bound for Columbus. Since then, I had no updates on my biological mother or Earl. Only a faded image of a broken disheveled woman standing in a messy kitchen, staring into space remained in my mind. That image grew dimmer by the day.

Crossing The Scioto

Whenever Aunt Elizabeth

wrapped her hands in rosary beads,

calling me into the den,

It was time to pray.

I prayed

not to be sent away.

I thought my days in summer paradise were numbered the morning Aunt Helen drove us across town towards Grandview Heights. Nervously biting my fingernails as we neared the Heights, I sat in the back seat while Aunt Helen hummed a tune. She parked at the *Grandview Outdoor Store* and told me to wait in the car. I tried to be patient, even tried reciting poems I had learned in school. Growing antsy, I couldn't wait any longer and left the car to find her. Before I could enter the store, out she came followed by a young male clerk pushing a shiny blue *Schwinn* bicycle with chrome fenders, mud flaps, and handlebar streamers. *Please let that magical ride be mine*, I prayed.

Wide-eyed, I held my breath, watching him load it into the trunk, as if I exhaled the entire dream would dissipate.

"Well," my aunt said to me once we were settled in the car. Her eyes sparkled, crinkling at the corners.

"Is it for me?" I dared to ask.

"All for you," she said.

The entire way home I dreamed of my first bicycle ride.

The next day, I peddled three miles to the *Scioto River.* Trekking through the thick weeds along the bank, I came upon the sacred serpentine Indian burial mounds hiding the secrets and bones of long dead conquerors. Not wanting to disturb the occupants, I continued exploring. In the thick wood, I retrieved buckeyes shedding tears and drifted closer to the water to comb the riverbank for skipping stones.

The dark water beckoned to me, reaching for the rocks in my hand. I wound up my throwing arm, and hurled flat stones across the dark river, watching each stone sink closer to my side of the bank as my arm grew tired and I searched for new entertainment. There, possibly 15 feet away, I spotted strips of shiny steel glinting in the afternoon sun. Then, arms groping for balance, eyes glinting steel danger of slick rails, I danced across a railroad trestle's oil-stained ties to a bullet-riddled *no trespassing* sign dangling on a chain blocking entrance to an abandoned limestone quarry. Standing on the precipice of an open pit, I gazed into a giant man-made crater strewn with ruins.

Grasping jagged rocks and broken roots, I started the descent to the devil's slide to the bottom. My boots were perfect: finding gouges in violated granite and holding my foot securely to the wall. I feared being skinned alive, though I had great confidence in the stickiness of my soles. As I dropped further into the hole, dust rose up around me, filling my nose and mouth with a chalky film.

The bottom of the quarry housed all types of interesting items. I found dozer treads, ripe for throwing like a frisbee, wove among engine parts, broken hoses, dead batteries, rusted cans, patches of oil, and knobby tires black as a mortal sin. Collectively, it all represented man's sins against nature.

Hours later, clinging to the roots of defiled earth and exhausted, I scaled the sides of the quarry. I never looked down; instead, I focused on my body snuggling razor-sharp stone that only would begin to dull through other millennia. Feeling the sting of cuts on my arms and legs, and seeing blood seeping through my T-shirt and jeans, kept me aware enough to push through to the top. My desire to live plus the sound of a speeding locomotive pulling train cars loaded with coal chased me up and over the rim. Sitting on the edge, I saw the train racing down the tracks, leaving its black cloud as a calling card. Years ago, those coal cars would have carried ore from the Carney Coal Company strip mine before its demise. The sound of a coal train rumbling across the Scioto River trestle was anything but peaceful; it sent shockwaves through the trembling earth, warning all living creatures that they were in peril.

Before I knew it, darkness was upon me. I grabbed my bike, raced home, and undressed in the basement. Scratches and deeper cuts lined my entire body. Knowing my aunt would demand to know my whereabouts if I asked for help tending to my injuries, I doctored my own wounds with Mercurochrome, applying it liberally to the deepest ones.

Aunt Helen had another surprise in store for me. After receiving the bicycle, I couldn't have imagined anything else would be as grand. The two of us arrived outside *Lane Avenue Shopping Center's Hobby Shop* one afternoon. I never heard of a hobby shop.

"Jerry, you'll enjoy this," she said while holding the door open for me.

I stepped through the doors and into another universe.

Turning in a slow circle, I tried to take it all in. My aunt, seeing my eyes light up while scanning assembled models in the display window, opened her purse and retrieved a crisp 5-dollar bill. I bought my first model kit that same day.

On the drive home, we worked out a solution to my lack of pocket money. Aunt Helen decided that I would receive a $2 weekly allowance and possibly a bonus for making bike errands to *Tremont Center* with her handwritten note to buy a pack of *Pall Mall* cigarettes. No regular chores were required for my weekly payment. I dreamed of all the models I could purchase with such fantastic sums of cash.

On rainy days, Jane found me hunched over my desk, inhaling *Testor's* glue and dipping a tiny brush in a thimble-sized model paint bottle. I grew obsessed with coloring wood and plastic fuselages, wings, landing gear, cockpits, guns, and ball turrets of bombers and fighter planes that defeated the Japanese, Nazis, and Communists.

One day, I thought to myself, *I'll fly a fighter plane of my own, or I'll join the army. I'll be braver than my dad ever was.* It would be years before my vow would be realized.

Convict World

Rows and rows of social outcasts,

seeking redemption for awful sins

I prayed hard to be good besides

my aunt.

We are all God's children,

she said.

My initiation into the criminal justice system was being patted down by security guards at the Ohio State Penitentiary; a massive stone fortress with iron-grate windows, guard towers, and cells that housed hundreds of Confederate prisoners during the Civil War.

"Now, Jerry," my aunt said, "some prisoners have no family to visit them. God loves all his children. We must show these poor sinners that they are loved by Him."

I was nervous initially, especially after the guards confiscated my crayons, declaring them "contraband" and inspecting the contents of Aunt Helen's purse. We were escorted behind convicts marching lockstep in silence across a vast yard to the prison's chapel. Aunt Helen walked us over to a small group of well-scrubbed white ladies in Sunday dresses and lace kerchief-covered coiffured hair.

"Warm afternoon, isn't it?" My aunt greeted the other women. They chatted amongst one another while I stared fascinated by the identically-dressed men following an officer across the gravel. Soon, Aunt Helen tugged my hand, leading me into the church. We climbed the cement stairs to the second floor, sitting in the lofty choir section. I peered over the edge. Inmates walking single- file filled into the pews below us. Following mass, the chaplain led us to a vestibule occupied by a dozen chirping multi-color parakeets landing on him and his guests' shoulders and hands that he called, "My Feathered Flock" as if he were St Francis Assisi. And as if trained by their keeper's quiet voice and gestures, the tiny birds returned to their cages.

Clean-shaven prisoners wearing starched denim shirts with I.D. stenciled numbers above pockets, pressed dungarees, and spit-polished shoes met us. Black men had shaved heads or short wooly hair white as cotton. Their white counterparts had burr haircuts, leathery skin, and *Howdy Doody* faces. These men reminded me of Appalachian folk: miners with stoop shoulders, missing teeth, and gnarled fingers, flimflam men and moonshiners I'd seen while hunting with Uncle Mutt in Wayne National Forest. The only difference between the two was their location, that and the prison garb that indicated where these men were allowed to go.

Curious about the lives of murderers and thieves I'd only seen in gangster movies, I interrogated convicts at every opportunity at the prison. Wide-eyed, I sat engrossed in their stories told in raspy voices. They boasted about jewelry heists, burglaries, bloody fights, and armed robberies. Inmates all proclaimed either their innocence and vowed revenge on those who framed them or acknowledged their crime as if they were admitting to taking a bite from a cookie. Those inmates made it clear their main regret was being caught for their felonious crimes. All were obsessed with counting down the days to their parole review and, hopefully, their freedom.

Sisters Of Charity

Learning to be contrite

is as much of an education

as history,

as math,

as reading,

as writing.

Lasting far beyond the four walls of the classroom.

My cousin Jim had been an altar boy at St. Agatha Elementary School before moving on to Upper Arlington High School. Yet, Jane was homeschooled by her mom in her early elementary years until she recovered from rheumatic fever and enrolled in a local public elementary school that had a nurse on duty which was not available at St. Agatha. However, Jane would be attending high school at Columbus's exclusive Catholic girls' institution, St. Mary of the Springs.

Given my aunt's preference for Catholic schools, it's no wonder she placed me at St. Agatha's where I was taught by the Sisters of Charity of Nazareth. I was assigned to a combined third and fourth-grade classroom. Unlike Dominican nuns at St. Bridget's and St. Christopher's that wore traditional black habits from head to toes, Sisters of Charity wore a white Kentucky sunbonnet one feared would tumble off in a sudden gust of wind.

Attending the school's mandatory morning mass, marching to the flag, and reciting the Pledge of Allegiance were tolerable. I even endured participating in rote learning, regurgitating forgotten history, dates, and words all while patiently waiting for recess time. However, my patience ran out once the school introduced a new dress code. I led a single-person revolt against their banning of combat boots and jeans, my preferred attire. The school imposed uniforms upon the student body: white shirts or blouses, blue cords or pleated shirts, checkered knee socks for girls or plain white socks for boys, and brown oxfords. When Aunt Helen held up my new school clothing, I refused to wear it. She just shook her head.

At school, I continued my protest. I splattered my history book with ink and boycotted learning. As a consequence of my defiance, Sister Mary Alda assigned me detention until I wrote a letter of apology and recited the Gettysburg Address. *I'm not going to go, and she can't make me, I thought to myself. If I run out of the school, she can't chase me down. I'll escape*.

At dismissal, I raced out of the classroom to make my escape. Instead of running into freedom, I ran into the arms of Sister St. Frances, the school's principal. Built like a Sumo wrestler, students called her *Diesel* for her ability to stop a kid in his tracks. Chagrined, I gritted my teeth and offered up my palms for my punishment. After a dozen ruler swats, she dragged me by my ear back to the desk I had vacated minutes before. Confident I'd outlast my jailers, I doodled for hours, sketching fighter planes, ships, and soldiers. Nuns supervised in shifts. Soon, the brightness of daytime became a pink sunset and then darkness. My stomach growled loudly. I imagined my family, eating together and wondering where I was. Weren't they worried about me? Maybe the police would find me. I crossed my fingers and hoped to be rescued from my jailers. Just when I thought I was about to faint from hunger, a nun brought me a cold baloney sandwich and a carton of milk. With a full belly, I dozed off still at my desk. At 9 p.m. I felt a tap on my shoulder. Aunt Helen grasped my things and led me out of the room without saying a word.

Thinking I was off the hook, I returned to school the next day dressed in my school uniform. I had won. I outlasted them. No way would I have to apologize to Sister Mary Alda or memorize and recite the Gettysburg address. I grinned.

After class, Diesel stopped me at the door. Nooooo! My plan had failed. Silently, I extended my hands toward her for my punishment. With my palms stinging from the ruler swats, I sullenly returned to my desk.

I was a prisoner again.

Again, Aunt Helen picked me up late and dropped me off at Mass the next morning. Three days of classroom arrest and baloney sandwiches took their toll. I surrendered, wrote a letter of apology, and stood in front of Sister Mary Alda's combined fourth and fifth-grade class and spoke like Mr. Lincoln. When the pastor passed out report cards, Sister Mary Alda winked at me. It was the first report card in three years– since Miss Zimmer tutored me– that I had more passing grades than failures.

When it came to exploring books, magazines, newspaper cartoons, and comic books at home, I focused more on the images than the words. Often I preferred the entertainment comics offered, ignoring my homework. Sometimes, I'd sloppily finish it before class started, but more often than not, I forgot to do it altogether. Quickly, I learned the consequences of academic failure or missing homework: teacher conferences, detention, and house arrest. All which interfered with my free time. My strategy at St. Agatha's was to keep my head down in the back row of a classroom. The fifth and sixth-grade teacher, Sister Mary Alfred was the tallest nun I have ever seen. She had superhuman eyesight and focused her attention on students whose native intelligence or achievement scores fell on the extreme ends of the bell curve. She favored straight-A students whose status was confirmed every six weeks when the pastor awarded them holy cards with their report cards. Students that misspelled words, talked in class, doodled in notebooks, forgot homework, failed a pop-quiz, hesitated to answer a question, were ordered to stand at the blackboard and write, "I will not talk in class,....not forget my homework," a hundred times. My nemesis was English and spelling. Diagramming a sentence was a jigsaw puzzle of nouns, pronouns, prepositions, verbs, adverbs, and dangling participles. Learning spelling and definitions was pure rote but incorporating a word into a declarative sentence was bothersome. Avoiding detention called for self-control, academic discipline, and home tutoring. Thanks to Jane's teaching, I managed to stay away from Sister Mary Alfred's radar. Of the report card's eleven subjects' averaged grades, most fell in the 71- 79 range which placed me in the sister's attention safety zone. The highest grade, 86, was in *penmanship* because I liked dipping an *Easterbrook* pen in the desk inkwell and writing in cursive. This grade followed low 80s in *arithmetic* and *spelling*– subjects that required memorization rather than processing. *Character Formation* was consistently checked, "Improvement is needed in exercises self-control." The nuns caught me giggling for flagellating during mass, shooting paper wads at a bully that pulled my chair out as I was sitting down for lunch, and other very serious infractions. They watched me like a hawk, ready to point out my next character flaw.

Over time, I understood the benefits of their relentless instruction. Their steadfast commitment to our education, conveyed through a mix of patience and discipline by selfless women in black robes, silly-looking sunbonnet hats, oversized rosaries and armed with rulers, showed pure belief in us– even the numbskulls and sinners. They truly were Sisters of Charity.

Moral Yardstick

Structure.

Many responsibilities

and religion.

The magic formula for energetic children

to become productive young adults.

Aunt Helen asserted the best yardstick for instilling morality in children was the catholic religion and catholic education. The path to citizenship, she believed, was learning the value of a dollar, frugality, self-reliance, a strong work ethic, and social development. She boxed me into the culture of true faith, mandating daily mass attendance with my class at St. Agatha's until I became a patrol boy.

On Sundays, she armed Jane, Jim, and me each with a copy of *St. Andrew's Missile* that I pretended to read. She forced me to take a paper route in the fall of fourth grade, encroaching on my nature excursions to the Scioto River and playing war with buddies. Every day after school, I came home to a heavy wired bundle of newspapers waiting at the foot of the Watson's driveway.

These bundles required a wire cutter before bursting open. I then hand-throw-folded every paper and stuffed them into a *Columbus Dispatch* canvas bike saddle bag or wire basket bolted to my handlebars. I maintained a monthly collection book containing the names and payments of customers. The longer and steeper the driveway the more likely no one answered the door or tipped the carrier. A carrier committed to abiding by the postman's creed: "deliver a dry paper in rain, sleet or snow." All I counted on were delinquent or missed collections, bike tumbles in unplowed and frozen streets, chasing down airborne pages of Sunday shove editions reporting *Buckeye* victories and defeats, cartoons, murders, and mayhem, economic booms and busts, discount sales, wedding announcements, and funerals. The only remedy for frozen hands, a flat tire, a broken link of a bike chain, and the twisted basket was filling manholes with all that bad news. My temporary respite from the work caused an avalanche of telephone complaints on the doctor's phone, which– to my and Aunt Helen's relief, ended my employment.

Aunt Helen's next imposed endeavor to enhance her ward's socialization as a *Cub Scout* member suffered a similar fate: A dishonorable discharge over refusing to weave potholders and make candles by melting old candle wax in a soup can with a string wick. Had the NRA offered kid programs in rifle marksmanship, archery, commando knife-throwing skills, or exploring caverns, I would have been their best member. Instead we melted candles or sat for long periods of time making potholders! Cub Scouts were *sissies*.

However, elated and relieved I felt over quitting a paper route and dropping Cub Scouts, I felt guilt for disappointing Aunt Helen, a volunteer *Den Mother* who purchased an expensive, non-refundable Cub Scouts uniform.

War Games

Gleaming porcelain figurines,

Chinese screens & urns,

made me restless.

I preferred rows of glass-encased

knives, lead soldiers, muskets,

swords and pistols.

Wished she left me

one of those damn heirlooms.

Sadly, I'd had to wait until the fifth grade to be eligible to earn a coveted place on St. Agatha's Elementary championship football team composed of 20 boys ranging from fifth to eighth grade. Unfortunately, I wasn't yet in fifth grade in 1949.

Meanwhile, I played war with neighborhood boys equipped with army surplus GI gear: entrenching tool, commando knife, field jacket, combat boots with strap buckles, leather gloves with wool inserts and sheepskin bomber coat, pilot cap with goggles, fatigue cap, and steel helmet with a plastic liner. My friend Dave had real Japanese rifles his dad brought back from the Pacific war we carried around the neighborhood. We would take turns killing ap snipers in trees, blasting Nazis in bushes and bunkers. My brothers in arms did not believe my cousin Jim had a coveted German *Mouser* rifle and bayonet, a bloodstain Nazi helmet with bullet holes, Swastika arm patches, and a German mortar in its original wooden case stored in our basement. His booty was given to him by his father's nephew, an Army Infantry captain that fought in the *Battle of the Bulge*; I could never show it off because Aunt Helen did not allow her children visitors.

Aunt Helen took me shopping for school clothes downtown at *J.C. Penney, Lazarus, Legget & Myers Army Surplus Store*. At *Gilbert's* discount shoes, I'd try on bargain dress shoe oxfords called, "seconds." I'd eye shiny leather for blemishes and flaws, inspect the shape of soles and heels, and stitch patterns for scuffs and defects. Make sure shoes matched and felt comfortable during a test walk. If it passed the test, I'd follow the salesman to the *Buster Brown X-ray* shoe machine.

"Step on the platform, son," he said in his voice reserved for adults.

"Yes, sir," I said. I stepped up, slipped my feet inside, and stared through the eye tube on top of the machine.

"Wiggle your toes for me, son."

I did as he asked, though I chafed at being called his son. I hoped my own father was not like this slick man.

"Now look here, ma'am," he pointed at the machine. "These shoes are perfect. His feet do not look cramped or deformed."

I saw my iridescent green toes wiggling for space, thinking what would happen if I was inside a giant x-ray chamber, what secret flaws would be revealed. *Would parts of my brain appear cramped or deformed too?* I wondered.

Not all errands I accompanied Aunt Helen on were for me. Sometimes she drove to antique shops in the country. I'd sit on my knees in the front seat, and gaze at pastures of lazy cattle, cornfields, and boulders of hay that appeared randomly dropped from the sky. And I'd shout at cats, ducks, and chickens scampering across roads, shoot crows and hawks with my finger and wave at farmers on tractors plowing fields. We paused at traffic stops by parks with war monuments of bronze engraved names of *Fallen Sons.* Wondering if the missing name was a hobo, scoundrel or John Doe buried in a pauper's field, I was eternally grateful that playmates never asked what happened to my mom and dad.

When we arrived at a building as old as its contents, Aunt Helen lead me by the hand into musty rooms of gleaming porcelain vases, crystal and sterling silver candelabras and pitchers, Ivory figurines, tall Chinese screens, faded paintings, posters, old calendars, pictures of President Franklin Roosevelt, Generals Grant, Sherman and Sheraton, marble garden birdbath sculptures of angels and saints, shelves of ornate clocks, painted China, pearl necklaces, gold earrings, and rows and rows of glass-encased antique knives, old coins, Confederate dollar bills, muskets, Civil War swords, and *Colt* pistols. Every shop carried its own special memorabilia. I learned more from watching Aunt Helen barter and check items for authenticity than I did during my entire year of school.

When she bought something, I was proud of her for getting the best possible deal. One time, I assisted her and the elderly antique shop owner load a cherry finish *triple-chime cable-driven grandfather floor clock* in the car trunk. The moon face clock protruding from the car's trunk looked like the head of a body in a coffin.

On the way home, we stopped at a *Dairy Queen*. Sitting in the booth, Aunt Helen saw our reflection on the glass of chocolate dripping down our chins, smiled, and said, "Jerry, let's make this a ritual."

Whether she meant the ice cream or the shopping, I didn't care. I loved every minute of the attention.

When we got home, she conscripted Jim to set the clock in the main hall by the den and charged me with the responsibility of standing on a stepstool and winding it with a special key to assuring it was always in working order and chimed on the hour.

Unlike the farm at New Lexington, chores were unheard of for most Upper Arlington children. The Watson's yard was maintained by a Filipino landscape crew that employed state-of-the-art power equipment. In the winter, Jim and I shoveled snow in the driveway. But all other winter work was hired out; handymen disassembled and stored summer awnings and installed storm windows.

Growing up in this house had its benefits. The main house was always immaculate. However, kids were made to feel like trespassers.

One day, I hoped to feel like I belonged.

While living in Arlington, I experienced fairly pleasant winters. Yet, one year brought a crazy snowstorm. In the evening, gathered in the family room, the radio blared, warning us of the impending winter storm of the century. We awoke on a white planet. Jane and I dressed in snowsuits and dragged sleds across knee-deep snow across the backyard, a vacant field, and *Lane Avenue.* We climbed the four-foot stone fence bordering the Scioto Country Club golf course, scaled snowy mountains, were buried in dunes, rode sleds down the steepest– the 9th hole– fairway and crashed in a frozen pond. We hurled snowballs at enemies and friends and pounded white powder off our coats and gloves. By mid-afternoon, icicles clinging to branches of wintry trees glistened, dripping tears of joy and sadness. The whispers of wind-struck willow trees drowned children's cries of free day's end and stolen breath.

Knowing our fun had come to an end, our shoulders sagged. Energy dissipated instantaneously. We walked slowly back toward our house, every step feeling like a mile. Our legs begged for relief. Our clothes were wet from sweat to the skin. We tied the sled rope tow lines around our waists and followed our tracks home. Complying with her mother's standard protocol, we entered the house through the basement entrance, shed our boots and outerwear, hobbled upstairs, and collapsed in our beds.

The next morning, sitting at my desk, gluing a propeller on the wing of a B-24 bomber when Jane suddenly raced to the window at top of third-floor stairs and said in a raised voice, "Jerry, Look! The Scioto Country Club's on fire!" She pointed out the window at black puffs rising in the blue sky above the white horizon in the vicinity of the clubhouse near the 9th green where we crashed our sleds.

"Let's go see," she said, pulling my arm after I jumped out of my chair.

"What on earth, children," echoed the voice of authority slowly ascending the stairs, reacting to the commotion.

"Shit," I muttered, knowing that Aunt Helen never came to my room because of her bad foot. She had no idea what I collected in that space. Jane knew since she occasionally washed her hair in my shower. Jim was a recluse and did not venture outside his own bedroom while home.

"My God!" Aunt Helen exclaimed, covering her mouth with her hand, breathing hard after the climb, wide eyes gazing at hell's flame and smoke darkening white sky. She unlocked her jaw, sniffed, turned aside, and said, "What's that smell?" Her eyes locked on a jar filled with yellow-brown water full of swimming guppies of which a few had met their demise, floating on the surface. Then she shifted her attention to a cardboard garter snake box hotel constructed at the foot of the bed, a posting "no vacancy" sign hung off the end. The snakes I had smuggled in during the last days of summer.

But I knew I was in deeper trouble when she stepped forward, her head hitting the landing gear of *B-17 Flying Fortress* flying a bombing mission escorted by *P-51 Mustang, Hellcat,* and gull-wing *Corsair* suspended from the ceiling by a thread. Her reaction was swift. She grasped a ruler off the desk, and swiped the sky with a ruler-sickle inflicting heavy casualties: wings, fuselages, cockpits, gun turrets, and landing gear came crashing down as if hit by Nazi ack-ack guns.

"Jerry- get these out of here immediately," she said, barely containing the rage in her eyes. She grabbed the jar and walked into the bathroom. She returned, the jar empty, the toilet filling back up behind her. Wanting to escape her before she could say any more, I grabbed the box of snakes that were returned to the wild. Yet, it took me weeks for the allied air force to regain control of the sky. These were my heirlooms, just as precious to me as hers were to her.

Thanksgiving Blood

Shooting a gun

for fun

and shooting a gun

for a kill

are not the same.

I didn't know any different.

My first Thanksgiving at the Watson's wasn't carved turkey and trimmings served on delicate China in candlelight and gleaming silver. It was a long ride to the Dr. Watson's family's country farm in Lancaster to eat rabbits and chicken. Clutching my tattered one-eyed teddy bear, I snuggled next to Cousin Jane, my boots poking her brother Jim sharing the back seat of Dr. Watson's *Buic*k. Aunt Helen slouched in front, as aloof as the stoic driver that never acknowledged my existence unless I was an anesthetized patient on a gurney at his hospital recovering from a suicidal leap off a garage roof, falling out of a tree or cliff trying to escape a world that already crashed. But what could I expect of a man not of my blood, stuck with the bills of raising a "wild child" that nobody wanted but Grandma Carney? I recalled feeling the softness of Jane's shoulder and the rhythm of her breathing. Even then, I knew she had a good heart when she called me, "Geraldine Roberta," fastidiously combed my hair, and wiped ice cream off my chin.

I accompanied the Watson male hunters at the Lancaster farm, while the girls and women cut vegetables, peeled potatoes, baked pecan, and pumpkin pies, prepared fried chicken, and set a table large enough to feed the Knights of the Roundtable. I cradled a stick in the crook of the arm like the doctor carried a 12-gauge *Remington* pump, traipsing snow-covered cow pastures abreast with adult and juvenile hunters. At fences, I followed gun-safety rules taught by Uncle Mutt at the Carney's country homestead: Slid imaginary shotgun under the fence's bottom wire, retrieved it at the other side, scanning the glistening landscape, then freeze at the sudden movement of leaves, the sight of bend in the blade of grass.

I was a true sportsman.

Only, I flinched at the report of exploding shotgun fire behind me. I followed the shooter to a patch of blood-splattered white snow, and watched him reach down, clutching a quivering cottontail rabbit by head and shoulder. I winced at each "crack" sound of snapped neck, each sound echoing louder than the first in my head.

I'd trail hunters behind the barn, witness the surgeon remove his leather gloves, retrieve a honed hunting knife, meticulously make a tiny incision in the rabbit's abdomen with the tip of the blade sharp as a scalpel, not cutting the flesh or muscle, slowly stretching the small opening, peeling skin with his bare hands like a sock, then gutting it in a wink pulling out entrails and parts unknown. My job was to bury the bloody pelts, severed heads, and rabbit's feet, covering up evidence of the crime scene. Though I held my breath, I still smell the stench and cringe at images of slick, skinned pink carcass tissue as thin as cellophane covering red and blue veins of remains fill my mind.

Once the deed was complete, my job was to carry the stinking meat by its hind legs to the hot, steaming kitchen. After several trips, Thanksgiving lost its appeal. I asked to be excused from the Thanksgiving meal, claiming illness.

The doctor merely grunted his acquiescence. The rest of the family was indifferent to my suffering except Jane. When she saw my eyes glinting, she followed me out of the room to the washroom as she rinsed the guilt and blood off my hands and affectionately combed my hair.

Stations Of The Cross

The cross,

of oppression,

of humility

of forgiveness

The next month I was introduced to the Watson family's Christmas Eve ritual. Aunt Helen ordered us, children, to fast, bathe, dress for mass and pile into the *Buick* at 11:00 p.m. sharp. The non-Catholic doctor only attended one mass a year, the Christmas Eve midnight mass at Columbus's St. Patrick's Cathedral. A parish headed by Father La Shay, the flamboyant Dominican Chaplain of Ohio Penitentiary, led the service. Aunt Helen insisted that we arrive early to sit close to the altar. We dashed to the second row, Jane and I using our winter coats to cushion the hardwood pews. She wore a special blue wool skirt and a red sweater that had embroidered green hollies, and a white lace veil like her mother's instead of a hat. I was groggy from playing all day. There were three altars with life-size marble statues: Jesus crucified on the cross over the center altar; Mary draped in shiny blue, gold, and scarlet layers of robes, arms extended with open palms standing at the left altar; St. Joseph at the altar right of his Savoir, arms hidden under a plain robe. Their eyes were mysteriously affixed to the vaulted ceiling.

Golden candelabras ornately framed Stations of the Cross and stained-glass windows reflecting the rays of a thousand candle flames and ceiling lights. Soon, choirboys dressed in scarlet cassocks with white shoulder capes, carrying hymn books, marched in and stood in front of the first pew on our side of the aisle. When the pipe organ warmed up, my attention went from the boys' slick-backed, parted hair to the people that had beaten us to the first-row pew. They looked remarkably identical in every feature except age, reminding me of *National Geographic Magazine* photographs illustrating man's evolutionary life stages. From grandparents to toddlers, each male and female figure had a thin face, beak nose, and jumbo ears. The only revealing differences among them were sex, skin texture, hair color, height, and thickness of glasses except for the youngest child who wore none.

I contrasted their physical appearance with those sitting beside me. Except for Jim's and Jane's prominent pointed noses inherited from their father, none of us looked related. My hair was black and straight, Jim's dark brown and curly, his dad's silvery gray, his mom's chestnut, and Jane's muddy blonde. From all general appearances, we may have come to church separately and ended up in the same pew by fate. I thought of my mother whom I hadn't seen in over three years, recalling that she had brown eyes like mine and red hair turning brown like rusted metal. I bit my lip, holding back tears. *What about my biological father?* I wondered. *Where was he now? Did we share hair color? Eye color?* I hoped that if we all sat in the same pew at mass, we'd look like the family in the first pew, clearly connected in more than one way.

I must have drifted to sleep because I felt Jane's elbow jabbing me awake.

"Don't worry, my mom didn't even notice," she whispered.

"Thanks," I grumbled, not at all excited to be awake again.

The church silence was broken by the sound of loud chimes. Everyone stood and gazed at the tall, handsome priest whose face I recognized from attending mass at the Ohio pen. Instead of a black cassock, he was draped in a silk white vestment embroidered with a gold cross-like seen in photographs or movies of crusaders. He was flanked by young adult priests dressed in black cassocks that briskly stepped up to the center altar to start the mass. These rituals seemed to take forever; I found myself nodding off again though I fought the enemy sleep waves valiantly. When it came time for the sermon, I waved the white flag, surrendering to unconsciousness.

Again, I suffered the onslaughts of Jane's elbow.

There was little doubt that the Dominican pastor knew how to hold parishioners' attention. Like Bishop Sheen, whom my aunt was addicted to watching on TV, he had a deep, resonating voice and gestured wildly as he spoke.

He focused the sermon on the sacrifice and suffering of Jesus. "The Son of God was born in a stable, slept in a manger, led a life of humility, only to be ridiculed, tortured, and left to die on a cross like a common thief," he almost shouted. "If the son could not escape such treatment, how can you expect to avoid any suffering?"

While the priest spoke, my eyes locked on the life-size crucified figure behind him. A crown of thorns penetrated his forehead like sharp edges of barbed wire, bright blood-red tears trickling down his pale face, staining his dark brown beard. Jesus's face was cast down on us sinners with a forlorn expression. I imagined him begging us to leave our pews and release him from his agony. His emaciated chest and bony ribs gleamed like polished marble. Blood dripped from a slit in his side where he was lanced by a Roman soldier and the flesh where his hands and feet were spiked to the cross. A cloth no larger than a ribbon covered his genitals.

The priest paused. I flicked my eyes back to him. He turned and pointed to where I was gazing. "See what *Jesus* did to redeem man's soul. How..., my fellow parishioners.., does that compare to your suffering and sacrifices?" he ranted.

Those words were meant for me and me alone.

The priest's words washed over me, bathing me in sudden awareness. When it came to my care, members of the Carney clan sacrificed much to keep me off the streets. It was apparent that Grandma Jessie Carney suffered and sacrificed the most. Like Jesus's suffering depicted in passages of the *stations of the cross,* Grandma Carney's motherly burdens began when she buried a stillborn infant. They did not stop when her beloved Jim, the promising first-born son, at 19 was crushed in the family's coal mine. Nor did her burdens end years later as her beloved, Grandpa Carney, succumbed to a sudden fatal heart attack brought on by years of hard work and hard times. She lost her adult children too– my Uncle Joe– and had to support his widow Evelyn and special needs daughter Anne. Even in old age, she cared for her intellectually challenged youngest son Ralph (my Uncle Mutt) who never left her side. On the side, she had a lingering heartache of mentally ill daughters, Kathleen and Wanda, who were committed to Ohio State mental institutions. She never saw her daughter Margaret and nor her two young children; their father, an ex-Navy WWII veteran, suddenly uprooted the family to California.

Then I was added to the mix.

She embraced the burden of becoming a caretaker for her orphaned grandson, never making me feel unloved or like the burden I knew I was. Her sacrifice was never clearer in my mind.

The loud sound of the pipe organ awakened me from a trance. The choir started singing *Silent Night*. I stood and glanced at Aunt Helen. She was teary-eyed. Uncle J.O. wearing a brown herringbone suit and blue bow tie projected his usual stoic surgeon's image, sporting two very dry eyes. Cousins Jim's and Jane's voices blended into the chorus. But I could only mimic the words until the choir started singing, *Noel*.

Suddenly, it was as if a great wind raced through the cathedral.

A gust brushed my face, tingling my spine. My eyelids were no longer heavy. Golden ornaments flashed and danced on the altars. Tiers of stained-glass cup candle offerings winked and floated toward me. I looked at St. Joseph. He remained calm. I turned to the Virgin Mary. Her opened palms were a receptacle for raindrops from heaven. Hundreds of voices and organ sounds passed through my body.

I no longer felt sleep-deprived, famished, and angry over being dragged to mass in the middle of the night, or my mind plagued by unanswered questions of my heritage or the drama and anticipation of opening more gift-wrapped boxes than I'd ever seen under a Christmas tree. As if magically transformed, I was erect and singing, almost screaming, blinded by strobing heavenly light. Forgetting, if only for a fleeting moment, that no one in the cathedral looked like me or shared my surname. Or haunted by the fading image of my mother's helpless gaze the last Logansport morning I saw her…., the whispers, "bad nerves," at my Grandview arrival.

Christmas Games

Christmas- a child's favorite day

even for an orphan

even for the black sheep

of any family

Christmas brings hope

And expectations to attend mass,

to stay awake.

I descended the stairs hearing Bing Crosby's voice singing, *White Christmas.* I entered the living room, inhaling the scent of pine. With wide eyes, I scanned branches laden with colored bulbs, red and silver bows, dangling angels and dripping tiers of tinsel and stars, and Christmas cards of Lincoln's face, sleepy eyes peeking at five-dollar bills suspended on a string over a pile of gifts wrapped in identical bargain green holly paper. Packages stacked around the tree on a fleece tree skirt soft as cotton. It was not as I expected. Not a listing, stubby tree with drooping limbs shedding dead needles, barely holding blinking lights and dark bulbs, strings of popcorn and water-colored walnut shells, and homemade paper rings crafted by Anne, Grandma Carney, and me. It looked worthy of being featured in a movie.

I eagerly started at the biggest gift-*Photoplay Football-* I had to plug in a wall socket and slide a diagram of a secret play revealed by a hidden light bulb like an x-ray to score a TD against Jane. Next was *Ping-Pong Basketball.* I pulled a lever to shoot a ping-pong ball through a wire hoop, and watched the ball dribble out-of-bounds, retrieved by my cousin Jim. We played *Chinese Checkers* with translucent blue, red, and green marbles, and opted to shuffle *Sorry* cards until bored. Finally, in the evening, I retreated to my third-floor sanctuary with my loot and read detailed instructions on how to assemble plastic models with the *Testor's* glue-like pieces of my shattered life.

I accepted my black sheep status in the Watson family hierarchy, a position affirmed by being shuffled off to Grandma Carney's in the summer months while my Watson cousins went to private youth camps and the Scioto country club swimming pool. I understood the house and managing the children was Aunt Helen's dominion. She purchased food, clothes, and supplies, and gave each child a weekly allowance and extra cash for chores. She oversaw landscape, house maintenance, and décor, and enforced house rules with no appeal. She fiercely enacted and enforced policies that discouraged friends from stepping foot in Watson's home. Sleepovers were reserved for my female cousins Mono, Betsey, and Susan, Aunt Helen's most prosperous siblings: Uncle Walter, a New Lexington business owner and inventor of heating devices, and Aunt Ruth who married a rising insurance executive.

Grandma Carney and Uncle Mutt were treated like country bumpkins and never invited to the Watson home. No photographs of the Perry County Carney Clan, including myself, were featured on their starched walls, only framed child portraits of Jim and Jane on the latter's bedroom wall. All rooms were immaculate except mine because the only adult that climbed the stairs to the third floor was Lizzy the maid. She dusted, vacuumed, cleaned my bathroom, changed linen, and replenished the dresser with folded jeans, socks, and shorts. Whenever my shoes or clothes were dirty from playing football, war, or sledding in the snow, I used the basement entrance door in the back of the house, stripped to shorts, and jogged three flights of stairs. The only recreation equipment at the house consisted of a ping-pong table in the basement. Thankfully, a neighbor gave me access to their basketball hoop.

All children in the Watson household had few rights and privileges. Jane and I watched TV in the living room, sitting Indian-style on a thick carpet next to the couch reserved for guests, under the watchful eyes of Venus de Milo. My favorite programs included *Superman, Lone Ranger, Cisco Kid, Nat King Cole, and The Liberace Show.* We were only allowed in the room if guests were not over.

Aunt Helen watched lectures by *Bishop Sheen* and the *Arthur Godfrey Show* on the small TV in the den. The doctor watched *John Cameron Swayze Camel News Caravan*, when he wasn't reading the newspaper, detective novels, or grading Ohio State Medical Board exams. When he was home, I retreated to my third-floor sanctuary to work on models or joined Jane sprawled on her four-poster bed, listening to *Hit Parade* over the radio, perused teenage girl fashions in *Seventeen,* eyeing graphics images in her father's medical books on how human life begins with zygote and ends by disease. Home alone, I sat in the doctor's red leather throne in the den, exploring encyclopedias A to Z, pretending to smoke a curved Sherlock Holmes-style pipe, whiffing acrid fumes of *Barking Dog* pipe tobacco.

My social obligations included behaving at mass, passing grades of Cs and Bs, and minding my manners in public. The only report cards I saw before enrolling at St. Agatha's were replete with F's and Ds. I tried to keep my head below the radar of nuns by staying invisible in class, learning by rote, Jane's tutoring. The only complete books I voluntarily read were *Hardy Boys.* I have no childhood memory of an adult reading to me while sitting on his or her lap. I rarely read textbooks, only scanned newspaper cartoons and comic book heroes. Outdoors was my province. Thanks to country roots and Uncle Mutt I could track, hunt, shoot, fish, plant vegetables, harvest corn, beans, and potatoes, harness and ride a plow horse, drive a mule wagon, mine coal, and raise a calf before starting school. Got my first taste of ether on the operating table at my Uncle J.O.'s hospital after my borrowed BB-gun shooting hand was impaled by a spike-size thorn diving for cover in a BB-gun battle. Side effects of self-prescribed activities like bleeding cuts sustained on slippery descents of limestone cliffs across from Scioto River, bike crashes riding no-handed, head-on tackles, and pile-on. A boyhood catapulting out the kitchen or basement door into an adventure.

1950

Roosevelt Park

Arming ourselves with homemade rubberband guns

with clothespin triggers

made from scrap wood

racing Schwinns to Roosevelt Park.

We were GI's attacking

the Reds.

Saving the world from

the threat of communism.

One rubber band at a time.

"Pack your jeans, socks, and *Fruit of the Looms,"* Aunt Helen said, informing me that Grandma Carney and Uncle Mutt moved across town after Uncle Joe died of a sudden heart attack. Their next abode was a peeling, simulated brink duplex that smelled like tar located on Mooberry Street in Eastside Columbus. The wallpaper was gloomier than an antique shop. My upstairs bedroom view was a dusty alley lined with telephone poles drooping wire, sagging garages, and overflowing trash cans. No Saturday poker night, TV movies, cartoons, popcorn, and soda pop refills at Uncle Joe's. I never saw my cousin Anne or Aunt Evelynn again. The adjacent duplex was vacant. The front porches faced a patch of dead grass, weeds, cracked sidewalks, and a cobblestone street littered with cigarette butts, candy wrappers, and old news. Grandma's Grandview furniture moved with them, including the rusty swing with deflated plastic cushions with images of wrinkled sunflowers, the tall oak, church shape radio blaring *Sky King, The Green Hornet, Romance of Helen Trent, and* Uncle Mutt's addictive program, *Lawrence Harvey with the News*.

I was sitting alone on the porch swing after a breakfast of hot *Cream of Wheat* sprinkled with brown sugar, playing the card game *war* against myself when a boy my age riding a bike suddenly jumped the curb, braked, skidded into the front steps, his body flying and landing at my feet. The bike had a bent handlebar with a missing handgrip, bald tires, no fenders, chain guard, or kickstand. The fallen rider stood and straightened a *Columbus Redbirds* ball cap. He had blond hair, powder-blue eyes, and a round face splattered with freckles. He wore a faded stripe T-shirt, rolled-up jeans, and high-top boots like mine.

He said, "What you playin?'"

"War," I said, turning over a joker and surrendering to myself.

"Phil, Phil Thomas," he said, extending an oil-stained hand. "I live around the corner."

"Jerry Wheeler," I said, reaching out and squeezing his hand hard like Uncle Mutt taught me, ("So people don't think you're a panty-waste.") I picked up the cards, motioned him to sit, made a quick shuffle, split the deck, and gave him his share.

His restless body rocked the swing, conveying he wasn't interested in sedentary activity. After a few matches, he jettisoned his cards on the seat, descended the porch, retrieved the bike, and adjusted the chain links that slipped off the teeth of the drive chain as if this was a common procedure. He said, "Come on."

"Don't have a bike," I replied. I assumed the bike stayed in the Watson's garage because I would return in the fall.

"So," he said, pointing at the handlebars.

"Gotta tell my grandma where I'm goin'."

"Roosevelt Park."

I mounted off-centered handlebars, quickly losing the confidence in surviving the ride on wobbly half-deflated tires, wheels missing spokes, weaving through trash and broken glass on a slick cobbled stone street lined with dwellings begging for another coat of paint or missing shingles.

Roosevelt Recreation Park was an island of hope and freedom. The Park encompassed a vast playground: a regulation-size basketball court with metal backboards and netted hoops, swing sets, a slide, monkey bars, tetherball pole, horseshoe pits, sand-covered baseball diamond, trampled grass football field, wooden picnic tables, and benches. Phil introduced me to a young, square-jaw ex-marine recreation leader kids called "Big Mack" because he had massive shoulders and arms and a bellowing drill sergeant voice. Each morning kids followed him like Pied Piper to a secured shed where he dragged out bulging duffle bags spilling a cornucopia of sports equipment: baseballs, footballs, bats mitts, horseshoes, basketballs, volleyballs, checkers, cards, and carrom board, distributed on "first-come-first-serve" basis and returned on the "honor system." A shortage of mitts compelled players at-bat to share gloves with opponents on the field.

Games started with the two best boys assigned team leaders. They faced off, selecting player by player until even the worst ball players, shamed by their draft number, joined a side. We played until we grew bored, tired, or hungry.

After ball games we traded bats and mitts for checkers, a carrom board and cue sticks or regulation iron horseshoes tossed at wetted-down pits designed for horseshoes to land in mud rather than rebound. This kept us from endangering the kids standing around.

At *Roosevelt Park* one's clothes, race or ethnicity did not matter. Being an athlete was akin to being a king. On the first day, I was treated as a serf and passed over several times. Whereas Phil was selected as one of the touch-football captains who took turns choosing players of respective teams. However, my initial low standing changed after I intercepted a pass, dashed to TD, and made other long runs for the winning side. To celebrate my move into royalty, I raced Phil to the park's 20-foot-high spiraling steel lighthouse observation tower in the middle of the park.

When we reached the top surrounded by bars and a safety wire, we hunkered down pretending we were carrying out the mission of WWII coast watchers scanning enemy territory with binoculars and signaling friendly battleships and cruisers the coordinates of Japanese bunkers hidden in the jungle.

"Phil, I see guns flashing out there," I shouted pointing at the imaginary sea horizon.

"Victory is ours."

"Victory!"

We played the game until the ex-marine shouted the order to return to our post on the ground; our cleated boots hitting steel ladder rungs orchestrating our retreat-like ricocheting machine gun bullets.

The next day Phil Thomas introduced me to his sidekicks that made up his football team. A natural leader, Phil played quarterback.

Tate Adams, a stocky short-tempered, pug-nose six-grader played center, loved to block, and ended each huddle clapping his hands with the words, "Fuckin Aaa!"

The team speedster was a Black kid named Herby Berryman, one of the few Blacks among park regulars. His moniker was "Blue Racer" because he had cobalt blue eyes and was the fastest ball carrier on the street or turf. Jacob "Skully" Skolnick played offensive end and safety. I rounded out the team by playing halfback and defensive linebacker.

I spent every glorious minute with my new team. It was heaven.

"Better Dead Than Red"

Red, blue, white

the colors of freedom

Black-

to me, a friend

to my grandma, an unequal

We were mates,

forming boarding parties

pillaging ships,

swinging on vines

Together.

Just two boys.

June 25, 1950. The "Eastside Gang" declared war on the North Korean commies. Pocket money earned running errands to the corner store, cashing in deposit bottles, two-bit lawn cutting jobs with Tate's rusted dull-bladed metal-wheeled hand-mower was spent buying GI equipment at local Army-Navy surplus stores. The first missions were camouflaging Grandma Carney's backyard fruit cellar door with Uncle Mutt's earth color paint and stockpile dust bombs filling lunch-size paper bags with alley dirt.

One hot late July afternoon after playing war I invited our ragtag squad of commandos into the kitchen for ice-cold lemonade while Grandma rolled cookie dough. After they departed, she called me aside, slowly placed the roller on the table, wiped her powdery hands on her apron, adjusted her glasses, focused tired eyes on mine, and said in a pejorative tone, "Jerry, what would you think if you came home and saw a colored lady sitting at our kitchen table drinking coffee with me?"

I reacted in jaw-dropping silence. Herby was an exceptional athlete. I looked up to him as all the boys in our gang did. I didn't know what she wanted me to say.

The rules of racism were meant to be inferred, not stated overtly. Until I met and played with Herby at *Roosevelt Park* and befriended him at home, socially Blacks were invisible. Starting at New Lexington, Logansport, Grandview, Upper Arlington, and Columbus Eastside. My interaction with people with black skin included glimpses of porters at train stations, housemaids cleaning Upper Arlington's fancy homes, waiters serving food, cooks in a busy kitchen behind a counter, and busboys cleaning tables at *Scioto Country Club* and *Columbus Athletic Club.* Once I saw black convicts marching lockstep with white inmates to the Ohio Pen's chapel, and serving breakfast to rich white ladies after mass. Other races I saw even less. I recognized Watson's Filipino lawn crew that operated giant rotary-blade power mowers and trimmed the shrubbery and trees.

Indeed, Blacks were invisible in my everyday life until playing at Roosevelt Park.

After I stood in front of my grandmother, dumbfounded, I became privy to how rampant racism was. During a scorching heat wave that summer, Skully came up with the idea of our gang cooling off at the *Jewish Community Center* swimming pool where he was a member, a bike ride away. We'd wear swimsuits under street clothes and follow him through the gate. But that idea was quickly rejected by his parents. This may have worked for white, non-JCC member gentiles but not kids with black skin.

Standing in front of my grandmother, I had never participated in a discussion about races. I found myself facing the Carney Clan matriarch who raised me most of my life without complaint, and with whom I had established an unbreakable bond. She looked me in the eye.

I looked away and considered the situation. Before this event, her most severe warning was, "for using the Lord your God's name in vain, Grandma will wash out your mouth with *Lava* soap." Now she was asking me to banish a loyal friend from our kitchen because he has black skin, skin that she referred to as "colored."

Mind you, none of the racial slurs like "colored," "negro," "nigger," "mackerel snapper," "dago," "yid," "kike," "spic," and "homo," were in my childhood lexicon. The word "queer" connoted odd, weird behavior, or untraditional fashion. Kids that demanded the return of bona fide traded or bartered items were called "Indian Givers." Aunt Helen scolded me for calling Watson's housemaid "Lizz," saying it was disrespectful. She taught me that alluding to one's race, ethnicity, and religion, beyond describing a trusted catholic friend as a "good catholic" was wrong.

Concentrating on not stammering, I took a deep breath, eyes locked on her eyes, I said, "Grandma, wouldn't bother me," then I dashed out the kitchen door, flying off the back porch and steps, nearly knocking down sheets sun-drying on the clothesline. I landed in the backyard, recovering my composure before sprinting off.

I reflected on what it would be like if orphans like me were grilled or ostracized by mates or strangers for their parentless status. Rather than being given a pass which thankfully was my experience, except the day the GI on a troop train asked why I was alone. Notwithstanding, the stigma of the disparaging term orphan was imprinted in my brain even if it was not branded on my forehead or worn on a patch like a Jewish star, or a disfiguring scar, physical deformity, or obvious mental deficiency. It was not visual like a wound; my other family members– Cousin Anne, Uncle Mutt, Aunt Wanda, my biological mother, Kathleen– and legless veterans hawking pencils in tin cups at street corners, mutilated casualty of war veterans who were given a wide berth by a passerby, or Black-faced kids denied entry at all-white swimming pools in Columbus the summer of 1950 all experienced discrimination based on appearances. This was my first experience, the first drip of a deluge of white privilege and its social consequences I'd witness, but it wouldn't be the last. Not for years to come.

The only days I wouldn't run around with the local gang were the days that the skies opened up, threatening to wash away small boys. During summer thunderstorms, we'd sit on the front porch swing, playing *Hearts, Fish,* or *War.* Grandma liked a hard rain because it watered down the dirt alley and washed away the soot-covered alley-side windows of the duplex. When lightning stopped, we'd remove our t-shirts, roll up our jeans, race barefoot to the curb, gather litter, construct a dam, watch water rise above the sidewalk, and flood dirt patch yard porch, until grandma enforced damage control with a scrunched face, and we destroyed the dam. Sometimes we chased handcrafted popsicle-stick-raft floating downstream until swallowed by a sewer. Regardless of the weather, attending Sunday mass was out of the question because the nearest catholic church was too far for grandma to walk and too close to bother with the bus.

The only time I saw paper money was the day grandma cashed her monthly widow's pension. Mutt and I walked three blocks to *Albert's* grocery store and bought essential baking ingredients, eggs, bacon, and toilet paper. He gave me cash to pay our monthly tab at the corner convenience store for emergency runs for milk, eggs, fruit, and over-the-counter medicine. Mutt earned cigarette and beer money from odd jobs painting houses and the like. But, unlike Uncle Joe who was rarely seen without a beer bottle in his hand and puffing a *White Owl* cigar before he passed, I never observed Uncle Mutt inebriated. Occasionally after a paint job, he took us to a street carnival and circus. I liked riding the Ferris wheel most. We gazed at a canopy of trees, shiny cars, and nice houses in the far distance. We saw a clown shot from a cannon, land on his feet wearing oversize shoes, dancing elephants, a lion tamer in a cage, ordering his charge to roll over and sit on a tool. We gazed and howled at skimpy-clad pretty girls riding rearing white horses, and others flying from trapezes caught by men in tights high above a safety net. Later, we were the only patrons at a honky tonk bar not dressed in western duds. Mutt talked about his dream of being a cowboy driving wild horses or longhorns on the open range, his future hope to visit California where his older sister Aunt Margaret and family moved. We listened to jukebox voices and strumming guitars of Hank Williams, Tex Ritter, and Sons of Pioneers blaring, *I Saw The Light*, *Rye Whiskey*, and *Tumbling Tumble Weeds,* as Uncle Mutt puffed *Camels* and seeped *Pabst Blue Ribbon* and I guzzled a Coke and munched on peanuts. But most mornings Mutt disappeared and then returned mid-afternoon, dejected, denouncing unions that he said denied him membership. He'd spread a newspaper on the kitchen table, and scan cartoons and employment pages that rarely advertised jobs for a third-grade dropout that walked with a limp and suffered a painful hernia.

But "Uncle Mutt's rules" stuck with me: "Always stroke in the same direction and clean the brush." He taught me the danger of dynamite, how to shoot a gun, harness a mule, carve a bow, plant vegetables in a row, and where to sit at the circus in case there is a fire. Other lessons too: "Never step twice in the same pile of shit." The only occasion he threatened to take a belt to me was when I shined him on after he told me to pick up my mess in the basement bunker. Lucky for me, grandma interceded when she saw him whip out his leather belt, fold it into a strap and raise it for a thrashing

She grabbed Mutt's arm, and said, "Don't you, ever, ever, lay a hand on my grandson."

He never attempted to hit me again, even though I was just an orphan.

According to the Oxford dictionary, "orphan" is a child whose parents are dead, have not been adopted, or prematurely separated from their parents. I wondered if the man named on my birth certificate was mean to my mother. Is that why he was never around. Maybe he never knew he had a son. No matter how much I eavesdropped, I never heard a whisper or a hint from Carney kin that I inherited any character flaw that explained Robert Vernon Wheeler's disappearance, or my mother's and Aunt Wanda's mental illness beyond, "it was a case of bad nerves." Likewise, I never heard a spoken word of disparagement toward Uncle Mutt who felt bad that he failed the army physical, dropped out of grade school, never qualified for a *Union* job, or acquired a skill that led to a steady job. The same nonjudgmental attitude applied to Cousin Anne whose eyes shone with puzzlement and shame when words didn't come, or hands did not obey. Meanwhile, living in Eastside Columbus, Grandma Carney continued to demonstrate country-woman steadfastness that raised nine children in the economic depression and two world wars, through tragedy, illness, death, and relentless unimaginable grief, without respite or complaint. Faults were to be forgiven and forgotten.

"Travelin' Light"

Visiting Grandma's twin brother

in Anna, Ohio

Might as well have been Egypt

Visiting the silver-haired man

took a long train ride

I dangled a rubber monkey on a string

out the window.

Visiting the man whose lungs were

burned by mustard gas

An atrocity of war.

We keep repeating the problem.

Holding our breaths and hoping for something different.

"Travelin' Light," Grandma said, packing our clothes in a carpetbag for a *Greyhound* trip to Anna, Ohio to visit her only living brother, Great Uncle James Webb.

The bus was a new upper decker with sliding windows that offered a panoramic view of Ohio's small towns growing bigger, highways lined with *Burma Shave* signs enticing riders to shift their attention from sweeping wheat fields, silos spilling corn, lazy cattle sharing pasture with grazing horses, foraging crows, vintage tractors hiding in weeds, sun-bleached barns with faded painted images of the American flag, and *Prince Albert* tobacco cans. To fight the tedium, I dangled a rubber monkey by its neck tied to a string out the window as Grandma snoozed.

Anna was a small country town Northwest of Columbus, smaller than New Lexington without the beauty of woods and rolling hills, or Columbus Eastside's recreation equipment at *Roosevelt Park* or *Champion* movie theater. But Anna, Ohio had a river too wide to skip rocks across, but cleaner and prettier than dirt-color Scioto. The Webb's lived on a narrow street lined with small frame houses with gravel driveways, and deep back yards planted with vegetable gardens. Both Grandma and her brother had silver hair, cocker spaniel faces, and Humpty Dumpty shape bodies. I deduced that Grandma's first-born son Jim-killed in the Carney Coal mine cave-in- was named after her brother Jim Webb, wounded in WWI. And may have influenced Aunt Helen to christen her son Jim in honor of her deceased brother Jim rather than her husband Dr. James O. Watson. Great Uncle Jim's wife looked and walked like a pigeon, treated us as intruders, and worse, consigned me to sleep on a sofa as hard as concrete. When I asked grandma why Great Uncle Jim always wore dark glasses and gasped for breath, she said his eyes and lungs were burned by mustard gas in the *Great War*. He spent months convalescing at the VA hospital and lived on a disability pension. Whenever we picked vegetables in his small backyard garden, I squinted at the sun's glare until my eyes stung and held my breath until my lungs burst.

On the first-morning excursion exploring Anna, Ohio's neighborhood, I came upon a group of boys my age, gathering at the corner *Sunoco* gas station. They rolled in on mud-caked bikes, executing wheelies, skidding stops, dismounting, dropping bikes on oil-stained asphalt, then raced inside to buy penny licorice sticks, jawbreakers, and ice-cold nickel bottles of *Royal Crown Cola*. Like most country boys of summer following spring harvest, they had burr haircuts, sunburned faces, and arms, wore tattered t-shirts, overalls or jeans, dirty *Converse* sneakers, and scarred clodhoppers. But what got my attention was seeing each rider armed with what appeared to be crude homemade wooden guns shoved in waist and shoulder belts strung with long rubber band ammo. Long guns were fashioned from 3-foot planks carved with notches that had a single rubber firing strip tacked at the front tip of the barrel and ran through each notch loaded with a single rubber band that stretched from the tip to the notch. The longest band of six rubber band bullets stretched from the barrel tip to the last notch closest to a rifle butt. Each notch held one rubber band bullet ready to be fired independently by lifting (pulling) the firing strip band. The most accurate, easy to aim, and reload gun was a single shot rubber band gun loaded with a bullet cut from a car tire innertube fired from a 24-inch-tip to a notch-barrel.

I asked one clearly experienced shooter, "How does it work?"

He enthusiastically said, "Watch and see," obliging a novice's request for a demonstration, and blasting a soup can down the road. The shooter used his free hand to pull a firing strip used as a trigger. The bullet velocity and power depended on the distance between the barrel tip to the notch, elasticity, and diameter of the rubber band itself, ranging from the smallest bike to the largest tractor size innertube. A warrior's arsenal included a single-shot derringer loaded with a bike tube bullet, a semi-automatic loaded with car tire bullets on a multi-notch >20in. plank with a firing strip, and a single-shot >36in. piece of wood (barrel), armed with a high trajectory "cannon round" cut from a tractor innertube.

During the visit, I happily volunteered to serve in Anna's rubber band gun army. The makeshift armory was a building marked for demolition near a lumber yard and junkyard. Basic training included scavenging scrap wood, sewing, carving notches, and salvaging inner tubes manufactured before WWII. Pre-war tubes had greater elasticity or lethal snap than war-era tubes. These were made of synthetic rubber. Cutting bands from inner tubes required industrial scissors and strong hands. After multi-firings, knots were tied in bands to preserve bullets' elasticity. Every morning we'd rendezvous at the gas station and choose sides. Designated foes wore arm bandannas.

Rubber Band War Rules banned striking the wounded. Anyone hit by a rubber band "bullet" was dead for the game. Even a shot in the arm, helmet, ball cap, foot, or shoe was mortal. Taking prisoners only occurred when a soldier was trapped and out of ammo. He must raise his hands and say, "I surrender." More often surrendering was induced by threatening to shoot an opponent in the face. I never surrendered but was shot in the back of the head trying to escape.

I learned many battle maneuvers and defense strategies in Anna which I later demonstrated to my Eastside compatriots. If ambushed riding bikes, we'd circle like covered wagons attacked by Indians, then blast the charging enemy from behind a wall of spokes. Chased, fighting in Eastside alleys or streets, we carried a cardboard shield with shooting slots. During assaults, we fired a rubber band cannon loaded with tractor tube rounds that knocked over bikes and cardboard forts. If the enemy thought one was out of ammo, cornered, and offered the option of surrender, we'd pull a loaded derringer hidden under a shirt or in a sock.

1950-1954

Head-Ons

I don't recall my padded shoulder

smashing into his thighs,

my helmet deflates his gut,

or locking my hands behind his legs,

hanging on for dear life.

I do recall gasping,

spitting blood,

feeling happier than I had ever been.

My summer obsession living on the Eastside was improving my agility, speed, strength, and endurance to make the St. Agatha's football team the first year of eligibility. Vying for a coveted slot on the Catholic Elementary School championship football team was a Herculean challenge for a fifth grader competing against older boys. I pushed myself so hard that I desperately sucked down air after running laps and calisthenics, my chest heaving under heavy plastic armor. A veteran lineman pulled me aside and said, "Impress the coach in head-ons. That's your best chance of making the team"

Drill after drill, I gave it my all until my bare knuckles were rubbed raw from crouching in a three-point stance. The wind picked up, blowing eerie warnings in my helmet's ear holes, but I continued as if each take were my last.

With my heart racing, I lined up facing a raging bull at least twice my weight. However, it wasn't the fear of excruciating physical pain, but the shame were I to be sent home in the coach's final fatal bleeding cut. The coach blew the whistle, and I jumped like a fawn sprinting for my life, the tiger of dismissal chasing me. My helmet crashed into the runner's padded thighs, my hands hanging on powerful, thrusting legs. He dragged my body; the earth smacked my face repeatedly, yet I hung on. Once the coach blew his whistle, I slowly rose, staggering and gasping for breath, spitting dirt, blood, and grass.

I blinked. I looked over at the coach. Across his leathery face, he wore a shit-eating-grin face mask.

Soon we switched, my team playing offense, the other team ordered to stop our advances.

As one of the smaller boys, my team decided I was to be a runner and tossed me the pigskin. Tucking it tightly into my shoulder, I catapulted toward charging hunchbacked fuckers. My ire was up–

these were the hulking monsters whom I faced against earlier. The whistle sounded. Suddenly, my body was lifted, propelled higher, and hurtled toward the ground in a suicidal dive.

This time I stayed glued to the ground for a few minutes before I could sit up. Stars danced behind my eyelids. Nausea threatened to bring up my lunch. My helmet lay ten feet away.

Once I opened my eyes, I searched everywhere for my fumbled ball and my pride, both could only be found by rejoining the play.

After three scrimmage trial practice days of punishment, we crowded around the coach as he read off the players who made the team in a monotone voice. "Adams, Arnette,...Smith, Wheeler... " he called out. Then he looked around, "I didn't say your name, better luck next year. For those I called, pick up your uniform before going home, you sissies."

With my school clothes stuffed in a bike basket, clad in St. Agatha Scarlet and Gold uniform, I raced home, shouting, "FUCKin Aaa! I made the team." I wanted the whole world to hear. Jerry Wheeler was someone.

Unfortunately, my excitement was short-lived. Fifth-grade players were relegated to scrimmage dummies and bench warmers until the St. Agatha Red Devils ran up game scores of another championship season. Happily, I earned a letter in the second season as an offensive guard and defensive linebacker in the two-platoon system. A great planner, Aunt Helen arranged for me to attend sports banquets with a surrogate father of a fellow teammate and my best friend, Bobby Bettendorf. Bob was destined to be a star halfback. Ironically, it was the loyal mothers cheering at the sidelines that deserved an invitation and a seat at banquets, but they were completely left out. MIA fathers who were too busy leading professional lives to sustain their families' social station by paying country club fees and children's college tuition were the only adults invited to attend with their sons.

I had no illusion of Uncle J. O. attending my game or award banquet. He never mixed family obligations with the business of medicine. He rarely interacted with his son, daughter, or spouse, and treated me as if I was invisible.

Come summer, I returned to Grandma Carney's eastside duplex for a summer of trash-hunting in dusty alleys, fighting rubber band gun wars at *Roosevelt Park*, running errands for pocket money, and watching anything playing on the big screen at the *Champion* theater. Watching newsreels, war, and cowboy movies back to back until I had seen everything playing. I anxiously awaited the next matinee double feature posted on the marquee. Visiting the theater was by far my favorite way to pass the time. Entering the cool darkness of the theater, I felt both rich and penniless. The actors were so glamorous. Leaving the theater, I was blinded by the sun as my eyes slowly adjusted to the brightness and reality returned.

Though Grandma Carney could barely provide for a growing boy, I never revered returning to a mansion's third-floor bedroom view even if it came with a view fit for a king: an expansive landscape, perfectly manicured trees, and vacant fields with wildflowers. Having maid service, a bike, allowance to buy models and supplies, a study full of books, and wild places to explore along the Scioto River trestle, caverns and quarries was still not enough.

I savored the hot summer evenings, hunched over by Grandma Carney's tall, vintaged wood radio, listening to adventures of *Sky King*, *Green Hornet* battling crime and evil, and to the latest bad news from the Korean war delivered on the *Lawrence Harvey Show*. Uncle Mutt puffed on his *Camels,* blowing smoke across the room as grandma quietly darned socks and patched my jeans. The sky darkened and lightning bugs lit up the night. This was home.

Teammate's Death

Loss

It twists and turns in the stomach

scraping and polking as it goes

Sometimes it travels North

bringing a fresh shower of tears.

I held them in.

I returned to my aunt's house just before the start of the school year. On the first day of school, the entire student body was abuzz with a terrible rumor. On the last day of summer, in 1952, Timothy O'Shaughnessy, age 12, while horseplaying, fell off the high diving board of the Upper Arlington Municipal swimming pool. He missed the water and crushed his skull on the edge of the deep end. Ironically, Tim was the youngest son of a prominent Irish Catholic family who owned a funeral service in Columbus.

Tim was tough. Like me, he also made the football team as a fifth grader and lettered in the sixth. As one of his closer teammates, his father asked me to be a pallbearer at his funeral mass.

Walking alongside his coffin, I wondered, *would his mother forever carry Tim's Immaculate Heart of Mary holy metal that he always wore, in her purse, like Grandma Carney carried her son Jim's army medals after he died in the family's coal mine? Is this what people do who lose their family?*

Following the funeral services, I changed into jeans, biked to the *Scioto R*iver, and skipped stones until my side hurt as if lanced by a spear of a ghost guarding the sacred serpentine Indians' burial mounds.

That fall, I was a meaner and stronger veteran lineman of St. Agatha's brutal two-platoon system: executing deadly tackles on defense, and carrying the ball, arising like Lazarus buried alive in human pile-on tackles, on offense. I earned a chance to demonstrate my newfound running skills when I intercepted a pass, dodged, and raced 40 yards to a winning TD. I soon proudly walked the halls donning a St. Agatha scarlet letterman's sweater with a gold chenille ***A*** patch with an embroidered football sewn on it and a chest dotted with "**I Like Ike**" campaign buttons passed out in the WASP Republican-dominated Upper Arlington community. Several rumors floated around the school. The first one, "Ike will end the war" seemed a given. The second was far more interesting: a new football coach was arriving next season.

His name was Jim Healy. Unlike his middle-aged predecessor that wore a business suit, tie, and dress shoes to practice and games, and never swore or dressed down players for missing tackles or fumbles. My new– Coach Healy, was a hot-head college student. He sported an *Ohio State* sweatsuit, low-cut football shoes, ***OSU*** ball cap, and cussed like a paratrooper at stragglers and pass receivers with "butterfingers." No one, even adults, escaped his temper. He faced off refs who dared penalize our side. Players whispered that he dropped out of *St. Charles Catholic Seminary* and was now a senior majoring in physical education.

One day during practice, he suddenly called the team together. Once all the boys stood around him in a huddle, he asked, "what do you call a guy who fucks his mother?"

Complete silence ensued. Boys looked at him wide-eyed, frozen like statues.

He paused, looking around a few times. Healy unequivocally loudly declared, "**A Motherfucker**! Now get back out there." We ran out, not even understanding what we saw.

Exhibition Scrimmaging

Hearing my name

over the speakers

validated me

I was someone.

1953 marked a new beginning at *St. Agatha's*. The school needed more space to fit the expanding student body. School classrooms and cafeteria were added, the church remodeled, and a new yellow school bus was purchased to transport students to games and field trips.

St. Agatha also scheduled two new exhibition football games that same year against *St. Vincent's Orphanage* at Columbus, and the *Chillicothe Veteran Hospital.* The first game fell on a fall Saturday afternoon of clouds chasing shadows across a freshly lime powder-marked field.

We walked on the field proudly. This was the season that St. Agatha parishioners issued players new state-of-the-art equipment and uniforms: shoulder pads, a shiny white plastic helmet with leather chin strap, scarlet uniform with jersey emblazoned with gold numbers, padded thigh and knee pants, matching red knee socks, football shoes with aluminum-tip cleats. My chest puffed up a bit; we looked sharp. Our team was accompanied by a few diehard mothers of players and the Sisters of Charity nuns, easily recognizable in their tall, white starched sunbonnet shape habits and shoulders draped in black capes as big as wings and long black robes that swept the cleat-stomped field.

In contrast, St. Vincent's players relied on donated mal-fitting, mud-stained, tattered uniforms, dented plastic, and dated leather helmets, worn sneakers, and oversized cleated shoes that reminded one of the boots worn by *Flippo the Circus Clown*. Exiting a drab building, St. Vincent players trotted to plank benches on the home team sideline overseen by three tiers of bleachers packed with castaway children. St. Agatha's team was relegated to the opposite visitors side where the only spectators were weeds, feral cats, and foraging crows. Also, St. Vincent's roster and fans included a few Black faces. It was the first team we had played that didn't consist of an all-white roster and crowd. Dressed in a white jersey, pressed Khakis, and sneakers, their white male coach appeared middle-aged, probably a priest or Catholic Brother, a male nun and ineligible to perform the sacred mass (sacrament). The home team's cheering section included a small group of Dominican nun teachers in full habit.

The nuns on each side anxiously paced their respective sidelines in laced high-top black shoes, lifting their long habits, exposing black support hose covering ankles; their hands nervously fingering oversize rosary beads marking respective team's forward progress or retreat in each movement of chains.

My first carry was a fluke: St. Vincent's short kick-off randomly bounced, landing in my unsuspecting arms at our 40-yard line.

I didn't miss a beat.

I tucked that pigskin in the shoulder, eyes madly searching for missing ally blockers and the best path forward for evading tacklers. Unfortunately their fast tacklers had no problem chasing down their flat-footed prey; I advanced five yards until gang-tackled by scrappy, Dickens Street urchins half my size. Spitting grass and dirt, I was lifted up by mates, slapping me on the back. I understood the tacklers' anger. They wanted to be someone as much as I did.

The next play, I raced around the end, sidestepped a linebacker, straight-armed another and advanced 10 yards until I encountered a goddam ragged midget racing diagonally across the field. Veering toward the sideline in the shadow of the goalposts, confident that I had the little shithead beat. Too late, I spotted out of the corner of my eye an outstretched dirty hand no bigger than a toddler's, touching the tip of my expensive *Spaulding* low cuts. Suddenly, I was tripping over sideline chains, flying topsy-turvy, and landing sprawled on the ground. No matter how surprised I was, I still was clutching the ball even while choking on lime dust and humiliation.

On the following play, no quarter was offered or received. I plowed over and dragged fellow orphans into the end zone. During the fourth quarter, we ended it. Well-armed gladiators fought against raggedy Christians battling in head-on encounters, trading forearm blows, high-low tackles, and piling-on after the ref blew the whistle. Our team was penalized for fighting and cussing but that didn't stop us. Bobby Bettendorf, our star halfback who lived in a mansion bigger than the Watson's, and I ran up the score until Coach "Mother Fucker" cleared the bench.

Bobby and I sat on the sidelines watching. "We got 'em," I said.

Bobby spit to the side. "They can't even take on our worst players."

It was obvious St. Vincent's motley team was ill-equipped and poorly coached. After the game, players shook hands and many visitors listened to youthful losers fantasizing about being adopted or rescued by long-lost kin. Older and wiser St. Vincent orphans stood around mutely listening. They knew an orphan's fate rarely traveled through the adoption station. One defeated child warrior hunched on a makeshift plank bench, staring at the cleat-torn earth said, "I've been waiting over ***six years***."

I was too ashamed to confess that my father was a counterfeit war hero I adopted from the cover of *Life* magazine, my mother in the insane asylum. I was effectively an orphan too. If it wasn't for being a lucky offspring of the Carney clan, I'd be wearing his ragged uniform. Thankfully, my peers never asked painful personal questions such as why no family member attended my football games or why I lived with kin. It wasn't a story that I wanted to share.

The exhibition game at *Chillicothe Veteran Hospital* stadium brought more spectators than we had ever seen. Hundreds of inpatient veterans, medical staff, and supporters for the opposing team were joined by a dozen of St. Agatha's die-hard parents and Sisters of Charity– recognizable by their tall, white starched sunbonnet shape habits, black winter capes, and long black robes. Our team's football cleats hitting the pavement echoed as loud as machine-gun fire once we disembarked the school bus, bringing three generations of war veterans to sharply turn our way. The service men sported many casualties of war. Veterans moved in two columns towards the field as if they were marching in sync. With cracked skin on their faces they shuffled along wearing unlaced combat boots mustering less national pride than the convicts, I had seen visiting Ohio Penitentiary with Aunt Helen or movie newsreels of German POWs. Young and old men with stoop shoulders, missing limbs, GI field jackets camouflaging scars of broken bodies jostling for space on tiers of wooden risers. Others barely were walking wounded: ex-grunts, sailors, airmen, and jarheads, cigarettes dangling from chapped lips, cold hands jammed in pockets of Navy P-coats, army olive-drab winter long coats, and faded fatigue jackets with winter inserts. The last group was assisted by nurses, wheelchair-bound amputees, their heads tilted like puppets with broken strings, hunched bodies bundled in Army blankets, wheelchairs parked on cinder track out of harm's way near the 50-yard line.

These were America's forgotten war heroes.

We saluted them as they moved past us.

Soon the predominately white spectators' attention was directed toward all-white rural grade school opponents practicing plays, and their respective entourage of supportive parents and coaches. Both teams shared the same bleacher side of the field. Of the estimated 200 patients, less than a dozen were black. They sat together in the middle of the bleachers as if they hadn't fought alongside their white peers overseas.

Since the fifth grade, I had played offensive guard, tackle, fullback, and defensive linebacker in the two-platoon system. 1953 was my final season at St. Agatha's. I never reflected on what was ahead in high school, college, or beyond. I knew if I failed any academic subject, I would be ineligible to play school sports and be grounded at home, yet beyond that, I didn't care much for academics. I treated most subjects like math, English, catechism, and geography, like chores. Some math problems required more time to figure out or tutoring by a nun or smarter student. My strategy was learning by rote. I cherished the Catholic system of canceling school on holy days, holidays and snow days. The rest of the time, I lived for recess breaks, and movies shown in class. At home, I perused *Look, Life, Colliers, National Geographic* subscribed by the Watsons and Encyclopedias in the den. I read *Guadalcanal Diary, Naked and the Dead,* and *Life Magazine WWII Photo Book published 1945.* Had I understood that the people who wrote for these magazines went to college, I may have been more focused in school. School wasn't only useful because it allowed me to play football; rather, playing football could allow me to attend school.

Before I could consider the future, I needed to resolve my current religious guilt surrounding my parentage. Ever since seeing the movie, *Boys Town*, I quivered and prayed to a God I didn't believe in that he wouldn't punish me for my father's sins or my conscious denial over not inquiring into my mother's mental status. Heretofore, I never felt the threat of family abandonment or rejection by blood. My armor was the Carney clan until I encountered the unadoptable children of St. Vincent's Orphanage. For better or worse, I accepted my orphan status regarding my absconding biological father, but not my mother's absence attributable to mental disease. Soon I would learn that the Carney armor was not bulletproof.

Meanwhile, the St. Agatha trophy case made room for another gold symbol of the gridiron championship season. And mother nature was about to throw me another curve: physiological and psychological changes affecting my breathing, mood swings, concentration, voluntary and involuntary reactions and thought patterns.

Smitten

Beautiful enough to start a war over

I set out to claim her,

crossing my fingers this would be no

Trojan War

Would I emerge like Menelaus?

Or would we depart together,

like Paris and Helen?

I was smitten the night I saw her at the cotillion, a parental mandated ritual teaching local eighth-graders social graces and dancing. This was a rite of passage mocked by spoiled early teens already into jitterbug and swing, petting, and French kissing. I agreed to participate on the condition that the trousers of my new required suit were pegged.

Her name was Carol Sievers. An eighth-grader at Upper Arlington Elementary, she lived next door to a St. Agatha classmate. She was one of the few potential students not staring at the floor. Dressed in red formal and matching heels, she scanned boys advancing like curious yearling colts across the dance floor. Her baby-blue eyes locked on mine as I moved forward taking the measure of her feminine attributes: long blonde hair that teased broad bare shoulders, a tan face, prominent cheekbones, flawless skin enhanced by a dab of blush and rouge, pug nose and rose-colored lipstick meticulously applied on perfect lips. Synchronizing our deep breathing, we pressed our bodies together, inhaled the aroma of sweet youth, and glided into space finding our orbit.

Henceforth, we were inseparable. Carol's casual attire consisted of a headband to control her locks, a sleeveless white blouse, white pleated short shorts, and low-cut tennis shoes. In the mornings, we biked to a local tennis court where she conducted tennis clinics, demonstrating that I was no match for her bullet serves, backhand smashes, and unreachable lobes. In the afternoons, the heat drove us to the pool. I was again humiliated watching her gracefully swim a lap, flip-turn like an Olympian, and move underwater at porpoise speed for a half-hour while I treaded water like a half-drowning walrus in the deep end. Afterward, we cruised to her house, necked on a basement recreation room couch, pretending to watch TV.

Carol's childhood was filled with health emergencies; she was born with a heart defect that required open-heart surgery and acquired juvenile diabetes. After my initiation , she casually unbuttoned her blouse, grasped my right hand, placed the tip of my index finger on top of the long, crayon-line-size scar, and slowly traced the scar down the middle of her chest with my index finger.

Straining not to wince, I schooled my face to be neutral. It was an act completely devoid of sexual arousal, yet one filled with trust.

Then, as casually as one straightens a collar, she adjusted her bra and re-buttoned the blouse.

I was acutely aware of her diabetes. After meals, Carol filled a syringe with life-saving insulin, loaded it in the spring-loaded device before injecting it into her thigh. She was diagnosed after she almost went into shock. Reflecting on the scale of my self-inflicted childhood wounds: crashing bikes, BB-gun fights, climbing trees and limestone cliffs, football injuries, compared to Carol's painful heart surgery, facing down a life-threatening disease, and her athletic achievements to stay strong, Uncle Mutt would have rightfully tagged "*Master Gerald,"* a certified "panty waste" for lady Carol.

Carol was my Helen of Troy.

1954-1956

California Exile

Ironed Levi's with no belt loops

stitching missing from the hip pockets,

and brightly colored socks that matched their shirts.

Every other boy

looking like James Dean.

I looked down on my powder blue shirt–

with fish and seashells

dancing over the surface,

my dark gray cotton cuffed pants,

and white Wigwam wool socks,

and knew I was a freak.

I was reliving my earlier night with Carol, replaying our evening spent gazing at shooting stars and exploding kaleidoscope shards in the July 4th night sky at Northam Park when Aunt Helen summoned me to the den. She was lying supine on the couch, her right foot propped on a pillow. Her eyes shifted from rosary beads lacing her hands and coiled in her lap to mine. Her face appeared pale, exposing wrinkles of aging and repressed family secrets. She wore a plain garden dress that had breast pockets with tiny bullet-shaped loops designed to hold golf Ts and a cigarette pack.

Without pause, in an unequivocal tone of a Mother Superior issuing an edict to a novice, she said, "Jerry, Auntie Helen needs a long rest. I have decided for you to live with Marg and Ernie in Long Beach, California. You are leaving early tomorrow morning with the Curtain brothers. Michael just completed his residency at the hospital. His younger brother, John arrived last week to drive them back home to Van Nuys. I enrolled you in an excellent Catholic high school run by the Holy Cross brothers. Pack your things in Jim's extra duffle bag."

She turned back to the TV. Our conversation was over

I was too shocked to backtalk. The aftershock of her words hurled me into the darkness of amnesia experienced on the journey from the Carney homestead to Logansport as a child. And it brought to mind the day a priest, without warning, yanked me out of St. Bridget's classroom and put me on a troop train bound to Columbus to be with my Grandma Carney. But in the context of my passed-around existence, taken in by the Watson's in the school year and then shuffled off to Grandma

Carney and Uncle Mutt in summers, was further evidence of Aunt Helen's ambivalence toward her nephew, and my status of a "black sheep."

I was all the more shocked since I had no inkling of being summarily expelled from the Watson family. I hadn't broken any rules. I was doing well in school.

What could have prompted this decision? To this day, I don't know. Leaving friends for a family I didn't know twisted my stomach in knots. Sadly, my models and other personals collected over the years would be abandoned like soldiers of a defeated army. Had we been at war, I wouldn't have blamed her. We were at peace. In a complete stupor, I packed my things and flung myself on my bed.

I awoke to the sound of revving diesel of 18-wheelers, my lungs burning from the fumes of *Mack* and *Peterbilt* behemoths drifting in a strange car's open windows. I was sitting in the backseat of a new *Plymouth Valiant,* next to a duffle bag stuffed with my past life. The car was following a trail of black smoke of tractor-trailers exiting an ornate toll booth on the Missouri side of *Chain of Rocks Bridge*. I watched the future Osteopathic healer and designated chauffeur retrieve change from his pocket to pay the toll. He turned his head toward the river and pointed out the window. He said, "Hey kid, we're crossing the great Mississippi."

Still in shock, I nodded.

He whistled a tune, paid the toll, and drove on.

As the smoke cleared, I saw a line of beautiful trees on the Illinois side of the gunmetal color river that was wider than the Scioto. Beyond the toll booth, we followed travelers past Missouri hills overlooking an amusement park and **Route 66** road signs– the pathway to St. Louis and scenic places out west.

Though I was a tiny bit excited to see the West, I never forgave Aunt Helen for the sudden expulsion and shock that prevented me from saying goodbye to Jane who I had established a strong emotional bond with and Grandma Carney who doted on her orphaned grandchild most of his life. Adding the pain of a thousand cuts, a 14 -year-old boy in puppy love torn away from his sweetheart, killed any emotion inside. I felt dead, indifferent, a shell of my former self. *Aunt Helen was positively cruel*, I thought to myself. *Someday I'd see them again, Aunt Helen be damned.*

Roadside Poetry

Poetry is everywhere

from the signs along a road

to the mysticisms of Native American rituals.

I reveled in the beauty.

But a drunken,

leering

asshole

cannot see it.

The landscape outside St. Louis reminded me of rural Ohio. We sped by country homes with elevated porches lined with rocking chairs occupied by old folk studying passersby, speeding past narrow roads displaying homegrown fruit and vegetable stands selling produce on the honor system. In the background, barns were painted with giant faded images of Old Glory, *Prince Albert*, *Bull Durham* smoking tobacco and *Camel, Chesterfield* cigarette packs, and accompanied various farm equipment: listing silos pouring corn, graveyards of rusted cultivators, vintage tractors with flat tires trapped in weeds, and sickle blades hewing ominous clouds. As far in the distance as I could see, appearing dropped from an endless sky, giant brick-shaped bundles of hay lay scattered in yellow fields. We passed fields with grazing mares eyeing rambunctious foals, and lazy cattle moseying toward quiet ponds and shade of ancient oak trees.

Soon, *Route 66* road sign poetry appeared:

CAR IN DITCH

DRIVER IN TREE

MOON WAS FULL

SO WAS HE

And more poetry:

Burma Shave;

HE SAW THE TRAIN

HE TRIED TO DUCK IT

KICKED FIRST THE GAS

THEN THE BUCKET

Burma Shave.

We checked into a motel room across the Oklahoma border. Michael, the quiet, short, parted brown hair future healer with a high forehead, dressed in an oxford shirt, blue seersucker pants, and *Topsiders* sat on the edge of the bed, studying a *Texaco* Road map. He pointed to the route to the *best steakhouse in Texas*-Amarillo, to make up for *White Castle* burgers that looked and tasted like burnt socks smothered in slimy onions. John, his younger skinny brother, had a flattop haircut. He wore a *UCLA* t-shirt, blue denim trousers, and penny loafers. He turned on *Zenith* TV which had a missing rabbit ear. Its finger-smudged screen flashed images of a technical snowstorm. The speakers sounded like attacking grackles. Sighing, he leaned back and opened a *Miller High Life* beer can with a church key, jettisoned his shoes, and fully stretched out on the bed.

The more he drank, the more he interrogated me about my cousin Jane. "Does she have a steady guy?" He said in a lascivious tone. "How old is she? How long have you lived with the Watsons?"

I answered his questions with as little detail as they deserved. Soon, he grew tired of my short, non-committal answers and grew quiet.

I lay in bed, reflecting on the strong bond formed with Jane over the years. We treated each other as siblings, teasing, playing games, sledding after winter storms, sharing secrets, and were openly affectionate. This was directly in contrast with her parents' and older brother's aloofness. I wondered, *was she forewarned of my exile? If not, how did she feel now? Did she protest? Or was she too scared to cross her tyrannical mom to ask?* I had observed Jane's transformation from a flat-chested girl to a young woman, touched her budding breasts playing doctor, and watched her taking a bubble bath as I sat on the edge of the tub discussing whatever popped up in our brains. I looked forward to snuggling in car rides or sitting in church pews. Whenever she got ready to wash her hair in my shower, she'd wrap her naked body in a bath towel exposing cleavage, scaling the stairs to the third-floor shower.

I'd heard Jane's mother boast to other mothers about Jane's "full figure" appearance and the importance of keeping her out of harm's way of "testosterone-driven" boys by sending her to *St. Mary of the Springs Catholic High School for Girls*. It was as if Jane was under house arrest so extreme that her brother Jim accompanied his sister to her Junior/ Senior prom. The milieu at home had nothing in common with radio or TV family images of *Ozzy & Harriet, Father Knows Best, Lassie,* or what I'd witnessed in adult recreational outings and sleep-overs with peers. The Watson's was a house of silence: soundless dinners, no horseplay, no pets– except a parakeet, and no discussion of politics, movies, books, or sports. No husband-wife or father-child interaction. I'd never seen Jim do anything with either parent. He was as quiet and aloof as his father. My singular experience with him was driving downtown joined by his visiting Notre Dame roommate to see *Shane.* The only total family rides were to thanksgiving dinners at Dr. Watson's kin in Lancaster, dining out, and attending Christmas midnight mass at St, Patrick's cathedral at Columbus. On a Notre Dame football weekend visiting cousin Jim, Jane and I shared a bed in a South Bend hotel room separate from her folks. Lying next to my attractive kissing cousin wearing summer pajamas, any latent adolescent sexual arousal succumbed to sleep deprivation from the long drive from Columbus and late-night participation in a raucous Notre Dame campus football pep rally.

"California's waiting, kid," said the future Dr., or as I had termed Dr. Asshole after he verbally lusted after my cousin Jane. He wasn't good enough for Jane, and I would help him see that if he ever said her name again. Thankfully he didn't.

I wondered what Carol Sievers was doing at that very moment. Would I ever see the graceful movement of her lithe legs racing across a tennis court and powerful arms execute a backhand smash out of reach of a stumbling opponent? Because of her hours of patient lessons, tennis had promise. Emerging after a dozen laps at the pool, I'd watch her remove her swim cap, shaking her head, wet hair spraying water drops in my face as I dried her broad shoulders, wrapped her body with a towel, listened to songbirds, children's laughter. We were inseparable before I left, so I felt optimistic about our relationship and our future. I was physically strong, self-confident, grounded, disciplined, mind-sharp and clear-thinking. Just like Carol, I brought boundless energy to a relationship that always seemed exciting. But then I remembered how, without warning, or iota of justification, I was being uprooted, sent two fucking thousand miles away by order of the oldest Carney sibling.

Did Grandma Carney know what was happening? I'd rather go back to her than live in California, a place I'd never been, with people I didn't remember..

I had anticipated joining my St. Agatha's classmates at St. Charles Boys Catholic High, Columbus Academy, or attending Upper Arlington Junior High School like my cousin Jim. *What would be my future,* I mused? It had been years since I'd seen my Aunt Marg, the Carney sibling reported closest to my mother growing up. She married Ernie Paskell, ex-Navy, WW II submariner, son of Perry County farmer, and father of cousins Bill and Marjean, ages 11 and 8 respectively whom I had not known since they moved to California from Columbus in the early 1950s. I had even forgotten what Aunt Marg looked like.

Sitting in the backseat heading west on *Route 66*, my eyes awakened to a wondrous strange planet described in the pamphlet I bought at the gas stop: a panorama of elongated mountains with flat summits a plane could land on. Others had ridges like dinosaur spines or tall, cone-shaped peaks puncturing low moving carpet clouds. Below these majestic mountains were foothills and rock slopes revealing pink feathery Apache plume, silvery leaved shrubs, pyramidal form black pine with evergreen needles, sage blooming red flowers, patches of bear grass with stalks bearing creamy white blooms, prickly pear pads luring the curious with pink, red and yellow flowers, and cholla cactus correctly described with *papery sheaths covering spines of segmented joints* ready to grope unwary hikers and critters.

Within hours, we bounced over unimproved roads in Indian territory dotted with teepee motels, ancient wigwams, single-pump *Texaco* gas stations, scattered livestock searching for water and shade, feral dogs scampering among adobe ruins, bullet-riddled hulks of pickups and cars, rusted auto parts, piles of old tires and coiled barbed wire. I envisioned Indians with war-painted faces, armed with *Winchesters*, carbines, bows that shot hunting-tip arrows fashioned from spent Yankee cartridges, hunkered behind boulders and sandstone cliffs awaiting the ambush of invading bluecoats and greedy white prospectors hunting for gold.

We arrived at Lupton, Arizona's landmark: *Chief Yellow Horse Trading Post*-the town's largest and most advertised trading post. Just past that were a row of ramshackle wooden structures and stands erected in a spacious cave carved in sandstone cliff at the base of the mesa. Like the first time I laid eyes on WWII souvenirs at Leo Myers Surplus Store, I was entranced by the spread: Japanese rifles, bayonets, GI steel helmets, and commando knives.

I examined handcrafted Osage and cedarwood hunting bows decorated with rattlesnake skin, buckskin, and feathers. I felt the sharpness of hunting knife blades, the weight of lethal stone tomahawks that crushed the skulls of 7th cavalry soldiers at *Little Big Horn*, and the glass smoothness of petrified wood.

I eyed the beauty and craftsmanship of Zuni and Navaho vintage turquoise sterling silver bracelets, rings, necklaces, and flashy squash blossoms.

Each had a story it begged me to learn.

A tall Native American woman in a buckskin dress, carrying a strange object approached as I replaced a knife in its sheath.

"What's that?" I said pointing at the object. My tone conveyed my ignorance; I was a pale face kid, born in Ohio, and seeing a live Indian for the first time. Her eyes glinted like black marble. Her face was dark red as rusted metal, skin creased by years of lost history surviving relentless droughts, sandstorms, and bone-chilling winter whiteouts. Her long braided jet-black hair with strains of gray fell over stooped shoulders like a heavy-duty rope, shoulders hunched from hours weaving intricate diamond patterns of proud clan heritage on blankets. Callused hands and arthritic bowed fingers sewed more beads than stars in the universe. Her traditional hide-clothing, sewn like previous generations of her tribe, had ornamental hand-made silver and precious stone embellishments. Her invaluable wares were displayed like booty of the conquerors.

"Dream Catcher," she said, holding up a wire hoop the circumference of a badminton racket. A spider web made of thread was attached to a wire ring bound by leather– a ring decorated with beads and strings of red feathers and suspended by lanyards. "You hang it at the head of your bed at night. The web catches your bad dreams, but lets the good dreams pass through."

I bought the largest hoop, hoping its web would shield the haunting nightmares.

We stopped at Needles-gateway to the Mojave Desert where I bought a pamphlet describing this mystical world. Most cars had a canvas water bag suspended on the hood ornament or bumper.

We drove through a paradise of retreating mountains without trails, an avalanche of stone that ended at the desert landscape with patches of grass shoots of cured, green and blue twine, shrubs that trapped lions, prickly pear cacti bearing fruit, claret cup cacti with yellow and red blossoms, giant daggers sprouting from the ground, one-arm saguaros directing traffic and critters helter-skelter, tall yucca blossoming creamy flowers, teddy bear cholla, ocotillo clip art, and skeletons hard as bone. Miles later, my eyes had adjusted to the iridescent gravel mound. I scanned a moonscape of scattered ruins: houses that looked like bombed bunkers, a toppled water tower pouring sand into a pond of broken glass near a gorge filled with rusted cans and tools too heavy to steal.

At Barstow, we entered an earthly phenomenon-California's revolutionary interstate highway system called a *freeway;* interconnected, multilane highways separated by wall barriers, going in opposite directions, and linked by bridge structures resembling trestles supported by concrete columns and steel struts. To our amazement, these roads extended for miles without gravel shoulders, stop signs, traffic lights or potholes. Racing vehicles dodged roadkill: carcasses of coyotes caught in headlights, dogs ejected from pickup beds, and flattened rapacious vultures with poor timing. Straggling cars were passed on both sides by impatient drivers. I wondered how these intersecting overpasses across the western sky sustained tons of moving cargo and hopes of people searching for a better life.

As we grew closer to our end goal, billboards popped up along the highway advertising different promised lands, each sporting "ranch family-friendly style homes" with grass carpet lawns, flowers, and modern conveniences. We entered feeder roads leading to small towns with used car lots selling vehicles with tampered odometers, advertising cars, "Only driven by school teachers. No credit; no problem."

Further into the city, we saw lot after lot with shiny convertibles, sedans, and pickups decorated with pennants, helium balloons, and American flags. These gave way to blocks and blocks of new of newly constructed strip malls with pawn shops selling guns, tools, jewelry of busted dreamers, flea markets overflowing with repossessed furniture, counterfeit antiques, racks of thrift shop clothes, caged puppies and kittens crying for rescue, A few sidewalks lined with orange trees waiting to be plucked by passersby.

It was too much.

The front seat passenger pointed at a road map of Long Beach.

He said, "Jerry, we must be close. East Stearns Street intersects with Studebaker Road. We're on that now."

Our car was going south, I sat in the back seat, staring the at plowed field adjacent to the east side of Studebaker Rd. The west side of the road held a housing development of ranch-style homes in various stages of construction. My heart raced when we braked at the stop sign, and I read names posted on the Street sign: "**E. Stearns St, & Studebaker Rd**." The driver turned left on E. Stearns St.

“OK, kid Look for **7127**,” the driver said. I looked, ready to get my first glimpse of my new life.

Tar Paper Stone Roofs

A travelling salesman after the war

Ernie never followed up on calls,

always on the move, conquering territory

spending commissions on gas and booze.

At home, he was obsessed with

outdoor chores

or gazing at the horizon through the windows.

He lived in California,

but carried an aversion to the ocean.

Stearns St. was a wide, three-lane, improved road lined with single-family wood-frame ranch-style track houses in various stages of completion. The property surrounding each house included an attached, two-car plywood door garage, and an asphalt driveway facing the street. Each unit's roof was flat and covered by stone pebbles spread over gooey tar. Less than ten feet separated the dwellings.

7127 Stearns St was on the north side, halfway down a long block that ended at a currently-dry flood control ditch adjacent to the plowed field. The field bordered the backyard properties of East Stearns St.'s north side including the residence of Ernie and Margaret Paskell.

I opened the car door, dragged the duffle bag over the curb, paused, and looked skyward. No hawks, crows, robins, or redwing blackbirds to be seen. Only a flock of seagulls soared overhead presumably toward the Pacific Ocean. I took a deep breath expecting to inhale the scent of salt air but only smelled oil fumes. My attention was quickly turned to the figures exiting the front door ten yards from the curb. The Curtain brothers stayed in the car parked in the street because the Paskell's driveway was occupied by a 1952 blue *Ford* station wagon and a red 1950 *Buick* Special.

I recognized my Aunt Marg who was short like Aunt Helen, her oldest sister, but more petite. With her curly jet-black hair, brown eyes, the Carney face, and a clear complexion, she could have been my mother. She wore a casual blouse, shorts, and flip-flops. She reintroduced her children, Bill and Marjean, three and six years younger respectively than me though I faintly remembered them from sleepovers back in Columbus. They both favored their dad: blond hair, pug nose, and pale blue eyes.

Her first words: "Honey, we've been anxiously waiting for you. I even stayed home from work." She warmly embraced me, motioning for her kids, Bill and Marjean, her husband, Ernie, and family pets, Toby a black cocker spaniel, and Marblelaid a furry yellow and orange cat.

I leaned down so Marjean, half my age, and height, with short brown hair and a face favoring her dad, could greet me with a friendly hug. She was dressed in a t-shirt, shorts, and was completely barefoot. Uncle Ernie extended his hand, his blue-green eyes sizing me up. He then clapped me on the shoulders, giving my bicep a squeeze.. His face was tan, hair yellow blond, and reminded me of the movie star Alan Ladd who played western hero, *Shane.*

But Ernie didn't have to wear elevator boots. He was six feet tall, even in his worn penny loafers coated in redwood sawdust. He sported an oil-stained peach color leisure suit; upper pockets filled with a pack of cigarettes and carpenter pencils. Not the picture of health, his beer belly protruded from his suit even though he stood hunched over. Later I learned his posture was permanently affected from months confined in a submersible boat packed with torpedoes, sweaty sailors, and a captain scanning the sea with a periscope searching for Japanese merchant ships. He was a man of the "silent service," awaiting the captain's authoritative voice commands: "Fire, One," "Fire, Two", "Down, Periscope" "Dive," "Dive." Bodies tossed, trapped in a steel tube pitching, and shaking from the impact of depth charges' exploding doom.

I followed my new guardians, cousins, and a scampering cocker spaniel across a small front yard of desert sand and patches of tiny, creeping round green leaves and my future nemesis, dichondra. The house interior included four small bedrooms, one converted to a TV room with a sleeper sofa, two side-by-side bathrooms, a combined living and dining room, and a kitchen with standard appliances and a built-in booth with a *Formica* table and *Naugahyde* seats. Instead of Watson's lush carpet, antique furniture, and expensive art, these floors were covered with Dark Oak Parquet except for the bathrooms and kitchen which were linoleum. The living room furniture included a credenza, dining room suite, red couch, coffee table, and pair of canvas butterfly chairs.

The only art was a photograph was a framed portrait of a father and young son wearing identical Navy sailor uniforms. My cousin Bill walked me to our shared room in the back of the house. He gave me the bottom bed of our bunk bed. The window side had a small desk and chair, offering a view of a barren backyard lined with anchored fence posts waiting for planks and bordering an endless plowed bean field. The bedroom furniture included a matching chest of drawers displaying stacks of comic books, model boats, toy trucks, a slinky, jar of marbles, and a pocket knife. But what got my attention more than the models, toys, or window view of the promised land: a gun rack accommodating a .22 *Remington* pump rifle, *Daisy Red Ryder* 1000-shot BB-gun, an archery bow, and leather quiver filled with arrows. A wall shelf contained a hunting knife in a holster, binoculars, and tube packets of copper BBs.

Bill and I were going to get along.

The first family conference was held in the living room within an hour of my arrival. Ernie did all the talking as his children and spouse, rolling their eyes and tucking their hands in their laps, sat passively on the couch. In total contrast to the Watson's and Grandma Carney's, I was now a member of a patriarchal family of working parents. Ernie's attention focused on the boys, as Marjean spent her leisure time with a children's family supervised by a stay-at-home mom. Every weekday morning before he left for his *Bostitch Stapler* salesman rounds, Ernie called Bill and me to the kitchen and outlined our daily chores. Aunt Marg, whom I always respectfully addressed as, "Aunt Marg," worked at a local bank. Perhaps due to my Carney blood and Ernie's in-law relationship and authoritarian tone, I couldn't call him "uncle." Ernie drove a 1952 *Ford* stick-shift station wagon loaded with supplies ranging from office staplers to heavy-duty devices employed in furniture upholstering. He'd make his calls on the road after scanning the morning edition of the *Los Angeles Times.* His work dress code was a white, short-sleeve shirt, thin tie, dress pants, and shined shoes. Before leaving the house, Ernie added his finishing touch: a plastic pocket insert to capture ballpoint pen leaks. He usually returned mid-afternoon.

Our morning work assignment required breathing the fumes of oil and turpentine and scrubbing off oil spots off a wood frame house constructed last year. It was new; however, it was a victim of an oil explosion at Shell Oil Co. Long Beach Signal Hill's oil storage facility. The blast spewed oil in Long Beach business and residential communities including ours. Our job involved three steps:

(1) wash away oil spots off white painted wood with turpentine rags;

(2) apply a coat of primer, and

(2) paint it with white paint.

According to Bill, insurance was paying for paint and labor. His dad paid us 65 cents an hour which provided me with money to buy a used bike and a *Benjamin .177 cal.* pellet rifle. Bill and I worked at a fast pace in the mornings, allowing us to inflate our hours and escape to the flood control ditch to hunt birds and plink around before his dad returned from business calls. Ernie changed into a leisure suit (work clothes), and moved the radio dial from music to the blaring voice of the L.A. Dodgers' Vince Scully announced, "balls," "strikes" and "fouls."

After completing the painting project, Bill started us on another project. Our mission was to unload, measure, and mark 3-foot lengths of each odd-size plank he bought at *Discount Lumber.* Thankfully, we were banned from touching the buzz saw. We hammered planks to 2 X 4 linked to redwood posts anchored in concrete. A standard fence used six-foot planks. But to save on the costs, Ernie constructed two levels of three-foot planks: a procedure more tedious and harder on amateur carpenters' fingers. We found no fucking relief from the sound of Vince Scully's voice describing the baseball world, the chilling sound of a screaming buzz saw, and Ernie's fucking orders: "Boys, more planks."

Land-Bound Ex-Submariner

Aunt Marg

Always working overtime

wishing her husband was a better provider

Forgiving him, for he spent

three years of submarine duty in the Navy–

months of no earth, sun or sky

hunting and torpedoing enemy ships

silent running in the bowels of the earth,

trapped in a steel coffin.

Praying not to be swallowed

by the sea for his sins.

I attributed Ernie's aversion to the sea, depicted in the TV episode of *Victory at Sea* showing the havoc of German Wolf packs torpedoing Allies' ships delivering supplies to Britain. He volunteered that only a few select sailors passed the rigorous training required of submariners. A recruit had to demonstrate that he was not subject to claustrophobia by being locked in a wall locker for eight hours without freaking out. He said cramping living quarters and long dangerous missions dictated that life on a submarine was informal regardless of rank. He described an incident when a cocky officer hoarded fruit and insisted enlisted crew members shave, call him *sir,* and salute him when they see him on shore leave. He related that one night in the middle of the Pacific this officer mysteriously disappeared on deck watch while the sub surfaced, charging batteries. No search was ordered by the captain: "Man overboard in rough waters." I read that submariners had the highest casualty death rate (twenty percent) of any military branch in WWII.

It was days until I saw the ocean and Navy ships across the Long Beach horizon. We picnicked at Long Beach. The surf was tamed by distant Long Beach-San Pedro breakwaters far from a sandy beach. Marjean made crude sandcastles with moats under her mother's watchful eyes, until flooded by the incoming tide. I dove in a waist-high wave, body tumbling, skinned alive by the power of the current dragging me on the sandy bottom, filling my nostrils and lungs, mouth regurgitating saltwater like vomit as my legs sprung to the surface, flailing arms crawling to the beach, riling sand crabs and spooking seabirds with spike peaks.

"Fuck this," I said, coughing and spitting foul-tasting water and pieces of green-purple plankton.

Swimming in a pool was hard enough. I'd stick to the dry sand.

Ernie supplemented his stapler salesman's job, loading a local milkman's milk truck at 5 p.m. Aunt Marg appeared to be the breadwinner. My cousins attended St, Bartholomew catholic school. I assumed Aunt Helen covered St. Anthony's tuition and my clothing needs. Ernie paid us on Saturdays. He had no idea how fast his son and new ward worked when he was on the road. After a few hours of accelerated painting, nailing, and laying flagstone or rocks bordering flower beds, we loaded our bikes with fishing gear and raced to the *Belmont Pier* or *Colorado Lagoon* a few miles from home.

A pier constructed of slick, slimy planks awash in fish guts, hosed water, pillaging seagulls diving between mussel coated pilings, and tangled fishing lines. We observed people of all ages fishing with various types of gear: crude bamboo poles, steel or graphite casting rods with expensive spinning, and conventional reels employing 20-pound line and treble hooks. A few adults wore fishing hats and vests, had fancy tackle boxes, water-filled coolers containing live bait (minnows); the day's catch dangling by a stringer from the rail in the muddy water below.

Further out and around the end of the pier, we spotted an old woman called the Shark Lady by locals. Her hair grew in patches of white bristles that sprouted out of her bald head like a sun-bleached rock. Her rhinestone eyes glinted like the ocean. She wore a bead necklace draped over a faded Hawaiian shirt with two missing coconut buttons, baggy jeans, and unlaced sneakers. Hunched over a waist-high cutting board, gripping a filet knife with her fingerless garden gloved hand, she sliced and hacked through the iron carcass of a three-foot shark for dear life. Her short body moved and rocked with the tide's swell; she triumphantly held her excised jaw to the sky, thanking the Lord before resting bloody hands on the edge of the cutting board.

The Girl At Colorado Lagoon

How did a huge creature

get to a small lagoon

surrounded by houses and roads.

Was it carried by a storm?

A castaway left to die?

Its only escape,

poachers stealing its bones.

At the pier, we'd use dead bait of a "throwback" or croaker. We'd catch, gut and filet enough fish to take home and fry for lunch. To expel the smell, we'd blast the evidence out the kitchen window with a portable fan. We counted on Ernie's cigarette smoke to mask the odor when he returned. Bored of fishing, we rode to *Colorado Lagoon,* flying down *Airplane Hill, the* steepest road grade in Long Beach. Next to the lagoon entrance was a strange house that had shuttered windows and a yard that had no flowers or palm trees. A car with hearse curtains was parked in the driveway. I asked a lifeguard if the house was haunted. He related that a teenage girl deathly allergic to sunlight lived there.

He said, "It's called *solar urticaria.* The slightest exposure made her skin burn, blister, and scar. It's incurable."

"Really?" I asked, believing every word.

He continued on once he saw my rapt face. Once, he said, he saw her leave the house, her face covered like a Muslim girl, led by a nurse to the car. She had to be completely covered up, or she'd shrivel up like a vampire.

I reflected on my tanned skin, earned by exploring nature, beach comb, gaze at warships and sailboats bellowing rainbow sails across a sunset ocean horizon in ribbon sea. It'd be like being a prisoner denied natural light, feeling the warmth of winter sun on my face. A life devoid of experiencing desert mountains, cactus flowers, eagles soaring in puffy clouds, changing the color of fall leaves, glistening snowbanks, the spring's first "robin blue" eggs, and female feeding a fledgling, was no life at all. Was the strange girl at *Colorado Lagoon* allowed outside at night to peek at the moon and stars? Was it the same as a child being warned that if she opens her eyes to see a three-dimensional world, she'd be stricken with a deadly disease?

Upon arriving at the lagoon, we gazed at the half-buried remains of a beached gray whale. We knelt in the sand and stroked a rib wider than our hand. The bone was grooved and felt like sandpaper. We brushed a spear-shaped tooth with our fingertips and poked an eye socket with our fists. I pondered how a giant creature got to a small lagoon surrounded by tacky houses. Was it carried by a storm; a castaway left to die? It's only escaped, poachers stealing its bones.

Will this happen to me, exiled from Aunt Helen's?

I imagined myself slowly dying from living outside of my natural environment. In reality, I was surviving fine, experiencing all kinds of firsts. My first frozen meal was a Mexican enchilada dinner served on a TV tray, watching Perry Como, Aunt Marg's favorite program. Eating off TV trays was a Paskell family ritual.

Aunt Marg's banking hours didn't allow her to cook every night. When she could cook, she took pride in preparing healthy meals: tossed salads, sumptuous fruit pies, and multilayer cakes made from scratch like Grandma Carney's. To ensure we had fruit and vegetables fresher and *cheaper* than *Safeway*, Saturdays at dawn, Bill and I helped Ernie remove *Bostitch Stapler* supplies from the *Ford* station wagon then drove to Long Beach Farmers' Market where an army of trucks filled with produce, hand-crafted leather goods, toys, and Mexican art, circled a park a block from the ocean. We encountered hunched old white men rolling small cannonballs in slow motion on a smooth clay carpet, lawn bowling, as other seniors sat on benches, smoking, and reading newspapers. We eyed Mexican families filling bins with tomatoes, lettuce, peaches, and every fruit and vegetable grown in California. Carrying a wicker basket, we walked slowly, stopped, husked, and inspected sweet corn, tickled navel oranges, and were dazzled by red and green peppers. We checked apples and pears for bruises, and a sign of worms, cried over onions, sampled slices of honey melon, and tasted salt air, then bargained like veterans.

After gathering our bounty, Bill and I punched pinatas with sticks, caught exploding candy, rushed to a leather mountain, and traced engraved snakes and cactus flowers on gun belts and holsters. We tried on sombreros and played Zapata and Poncho Villa until a little girl with coal-black hair, olive eyes, in a white lace cotton dress and sandals, standing on a crate, dangled a hand-carved puppet pistoleer, held us up and shot us full of holes.

Ernie rarely cooked or engaged in domestic duties. We kids bused, washed, dried dishes, dusted, and vacuumed. An elderly lady was hired to iron clothes. In contrast to Watson's world of books, and glitzy magazines like *Life, Colliers, Saturday Evening Post, and National Geographic* the only magazines found on the coffee table were *Popular Mechanic*, and *Photoplay.* Unlike Aunt Helen's dinners served on *Spode* china, including bread and butter plates, crystal glasses, linen napkins, and silver utensils set on fancy placemats at glass top wrought iron table, the Paskells' favored eating on TV trays, watching TV, or at kitchen booth. I soon learned the perils of the latter for their younger child.

Corporal punishment was not in Carney's blood. The nuns' answer to backtalk was ruler-swats on an open palm. Holy Cross brothers brooked no insolence, but their punishments came with warnings and usually followed my conscious breaking of rules.

I never witnessed the physical abuse of a child by an adult, yet friends relayed stories of beatings by parents for disobedience or corporal punishment administered by school principals for incorrigible behavior at school. For me, my physical punishments were limited to nuns' ruler swats on palms for farting at mass or harassing peers. Once, Uncle Mutt whipped out his belt for back talking to him but was stopped by grandma's harsh words. Only once Aunt Helen threatened to switch me for violating her commandment of "punctuality at dinner" with the doctor, but she stopped before making contact with my bare behind.

To my shock, I witnessed abuse during a family dinner in my initial weeks. Bill and I sat across from Ernie and Marjean at the kitchen booth. Aunt Marg sat down at the end of the table, the side closest to the refrigerator. She had just finished serving bowls of ice cream dessert after the main meal.

Lowering his brow, eyes riveting on the bowel as if a fly just landed on his scoop of *Borden's* vanilla ice cream, Ernie's face flushed beet-red (like Sister Mary Alda's complexion transformed, striking a student with a ruler), as he suddenly snapped, "git the damn chocolate syrup." With no warning, he turned toward Marjean, then smacked her face so hard the blow knocked her out of the booth, her body landing on the floor by her mother's feet. A partial imprint of his hand appeared on her innocent cheek.

She sucked in air and cried out. Aunt Marg instantly jumped out of her chair, lifted her sobbing child by the arms, and whisked her out of the room. Bill and I froze, not knowing what to do, then slowly slid out of the booth, and disappeared. We both feared that we'd be next.

Suddenly, I understood why Aunt Marg never left Marjean at home alone with her dad. I vowed to stay far away from his reach.

Rights Of Passage

Boys will be boys

they all say

What does that mean?

Aggressive?

Loud?

Independent?

Maybe they should say,

Boys should be raised to be men.

In Ohio at Grandma Carney's, I was hyper-energetic, hot-tempered, aggressive, and highly independent, loved sports and nature. I had an aversion to boundaries and boredom. I learned to craft rubber band guns and slingshots from rural kids that could not afford BB rifles or fancy cap guns and had no movie theaters to sneak into. Their playground was corn and wheat fields, barns, and silos. In Eastside Columbus, *Roosevelt Park* was the neighborhood rendezvous and center for working-class kids testing their talent and meeting new friends. My life in Californa was very different from any of my previous lives.

In Long Beach, by mid-August, earnings from painting and hammering paid for a *Winchester Model 69* bolt-action repeating rifle with a ten-shot clip, walnut stock, *Weaver* scope, and leather sling. Like Aunt Marg's male siblings, Ernie was raised in rural New Lexington where boys learned to shoot before they could read. Bill inherited his dad's *Remington* .22 pump. Ernie insisted we pass muster under his supervision at the local shooting range including proper cleaning and oiling firearms before he trusted us to plink alone at the flood control ditch down the road. First rule: never store or transport a loaded firearm. No round in the chamber when carried from house to the flood control. He prohibited pistols because a short barrel hinders the eye from quickly tracking a line of sight. *Winchester* Long rifle copper-coated *Super X* hollow point.22 ammo was expensive and only sold to an adult. Realizing bullets travel hundreds of yards in space and ricochet dangerously, we would sit on the concrete wall flanking the sandy bottom, blasting tin cans and bottles. Jackrabbits were elusive in the bean field and flying crows and seagulls had an invisible shield. Due to .22 ammo cost, we plinked more with the BB and pellet rifles, we occasionally aimed out the bedroom window and shot sparrows on the telephone wire. The .77 lead pellet was lethal. A sparrow was dead before it hit the ground. But pumping a pellet rifle with compressed air was harder than loading a musket. Eventually, the sibling rivalry in marksmanship waned as we examined gapping exits wounds and blood-splattered feathers of tiny birds held in the palm of our hands.

In addition to plinking, pier fishing, and relaxing at Colorado Lagoon, between chores Bill and I rode our bikes two miles from home to Bellflower Park to play touch football. Having played four years on St. Agatha's championship team, I was the chief organizer. Bellflower Park was a sanctuary for neighborhood misfits who never participated in their school's organized sports. No spectators or benchwarmers, Captains took turns choosing players according to talent. Boundaries were stumps, ball caps, and street shoes. Our Playbook was diagrams drawn on the grass. We clashed on an equal playing field. Cleat wounds were treated with Band-Aids and Mercurochrome. The star pass receiver's dad was an undertaker. Whenever he dropped a pass, he fell to the ground and played dead. My friend Smitty, too young to drive, worked on car engines blindfolded and threw bullet passes with greasy hands.

Time crawled by. Mid-August came and went, yet I received no calls, no letters from Ohio. Not one word from Aunt Helen, Grandma Carney, my cousin Jane or Carol Sievers, my dream girl. Aunt Marg never mentioned her siblings or her pre-California life. On her days off, she worked in the yard planting flowers and on weekends enjoyed packing us in the station wagon and going to the world's largest drive-in movie. At home, she was addicted to watching Fred Astaire and Ginger Rogers mastering the dance floor in old TV movies, listening to the soothing voice of Perry Como bring calmness after another hectic bank workday and the noise of Ernie's never-ending home improvement projects.

Sweet Reunion

Young love: A blessing

and a curse,

one that results in broken hearts.

I was emptying concrete into a pail from a rented cement mixer when Aunt Marg ran out of the house and said I had a phone call.

Thinking the caller was a Bellflower Park football buddy, I said, "Hello."

"Hi, Jerry, it's me." Before the next word registered in my brain, Carol Sievers's soft voice continued, "I got your phone number from your Aunt Helen. Guess where I am?"

Before I could catch my breath, she said, "Pasadena, with my family visiting relatives. Wanna go with us to San Diego Zoo?"

Suddenly, days of depression over my California exile and severed passionate companionship washed away. The next morning, we shared the back seat of her parent's *Buick Riviera* with her 10-year-old brother, Roger. Carol and I sat shoulder-to-shoulder, holding hands. Her copper-tone skin accentuated youth's natural beauty. Her blonde hair was cropped for summer freedom. One look at her blue eyes melted away all my sad thoughts of our separation. She wore her typical white sleeveless blouse, matching shorts, and blue low-cut sneakers. Her parents always welcomed my presence and trusted us alone in their basement recreation room. Touring the zoo by ourselves, we kissed and embraced in the shade of palm trees, eyed by caged tigers, lions, wolves, exotic birds, and snakes encased in glass. Riding back to Long Beach, I wanted to suspend time. Carol had only two days left in California.

Aunt Marg met us at the driveway upon our return. She asked if Carol could return the next day for a special lunch. I expected sandwiches served on TV trays or in a crowded kitchen booth. Much to my surprise, the following morning Aunt Marg called in sick. We attacked clutter and spiders, raced a Hoover, and polished wood. She ordered Ernie to connect the *Duncan Fife* leaf to the mahogany dining room table, and transplant flowers from the garden to the table before he went to work.

She brought out Grandma Carney's china, polished silver, and beeswax candles. We ate homemade rolls from bread-and-butter plates and drank from crystal glasses. After apple pie alamode, my aunt shooed Bill and Marjean away and disappeared. Carol and I danced to the *Kismet* musical and Sinatra then kissed goodbye.

It was the best gift Aunt Marg would ever give me.

Nothing of note happened after Carol left. I was despondent. On the last day of summer Bill and I tried for one last adventure. We rode the bus to downtown Long Beach, watched war movies and drunken sailors groping girls, gorged on sugar-caked donuts, and drank *Orange Julius* next door. We raced across the street to the *Pike Amusement Park:* flew a neon rocket ship to the moon, surfed a roller coaster in a gleaming ocean of battleships and cruisers. After we landed, we sneaked into tents, saw a sailor kiss a two-headed woman, touched a snake-skin boy, and ogled belly dancers with Marines until chased out by muscular Shore Patrol sailors swinging a baton. We spent our last quarter, strapped into gravity boots in the middle of three concentric rings in a contraption called a *Gyrotron*. Using the movement of our arms and legs were spun head over heels, sideways, executing cartwheels with no hint of dizziness. Broke, we thumbed home, swearing when we grew up, we'd join the Navy and see the fucking world.

I took my first token bus ride to *St. Anthony Boys High School* on Seventh Street. A fat woman driver chewing gum gave me 20 washers for $5. I dropped one in the token box. She pulled a lever as I watched it tumble down a chute, swallowed like my former life in Ohio. I clutched my GI backpack of supplies, grabbed a handgrip, and gazed at passengers from Mars. The boys wore brightly colored shirts that matched their socks, *Levi's* with no belt loops, missing leather *Levi's* patch, rolled cuffs, and stitching was missing from hip pockets. Their cordovans gleamed like marble and had a peculiar snapping device instead of laces. A few wore a *Wilson High School* red-colored letterman's jacket with **W** embroidered with a sports symbol in the middle on the chest and a numeral shoulder patch indicating graduation year. Their flat-top hair was slicked back on the sides, almost covering the ears. The guy rocking next to me looked like James Dean. I looked down at my powder blue shirt decorated with fish and seashells, gray cotton cuffed pants, polished oxfords, and white cotton socks, and knew I was a freak.

I scanned girls that had bleached ponytails and golden faces, wore push-up bras under tight fuzzy angora sweaters, and loud-colored hooped skirts that forced them to cross their legs and wrap their arms around their knees to prevent the skirt from flopping up and exposing their thighs and panties. Their legs and new saddle shoes bounced in unison to the rhythm of the bus's dips, jerks, and accelerations on *Pacific Coast Highway*. Their hair swayed between swishing doors as they exited at Wilson High, while I continued a 3-mile lonely ride to a stucco adobe tile roof fort guarded by the Holy Cross Brothers fingering pea-size rosary-bead belts tied around black cassocks. I scanned the students outside the school.

There was no sign of one damn girl or the ocean. Aunt Helen's controlling reach extended to California.

Once other students took a gander at my two-bit barber-college-flattop shaped like a Yankee Civil War cap, shirt with wildlife images befitting a third-grader, laced oxfords, and white socks, they called me "rank" and "rum dumb." While I hadn't heard these terms before, I knew what they meant. I got quite a vocabulary lesson that day.

When I got home, Aunt Marg asked what I learned on my first day at the new school,

I said, "A new word: bitchin'."

She raised her eyebrows at me, but didn't say a word.

I had two lay teachers: Mr. Navarro, a tall thin curly white hair Italian who taught Latin– a subject I dreaded, and Mr. MacDonald, the English teacher built like a retired wrestler. According to rumor, he was a former professional defensive linebacker and a man of few words. He was old enough to be my father if I had one. The English classroom door was unique. It wasn't a thick, ornate solid oak like other classrooms, but plain plywood. In the third week, he distributed a list of poems and told us to select one and be prepared to recite it by memory in the next class. The following day he summoned Tony Blakely, a lanky kid to the front of the room. MacDonald asked Tony the name of his poem.

Tony said, "The Charge of the Light Brigade."

When Tony stammered and froze after saying, "a half-league forward," MacDonald left his desk, walked over to Tony, lifted him by the arms, carried him a few steps forward then hurled him through the plywood door, creating a human-sized hole in the panel like seen in a cartoon. I understood why the English class door wasn't solid oak and gasped in relief that I had chosen Dickinson over Tennyson.

The student body was composed of mostly middle-class white Catholics and a few non-Catholics expelled from public schools. In contrast to all-white St Agatha's and Upper Arlington public schools, St. Anthony's student body included one Black and one Hispanic boy. Fights over car club members' parking spots, personal grudges, jocks vs. non-jocks, and hazing new arrivals were rare and resolved at the bus stop in front of the school unless a Holy Cross Brother intervened and ordered the contestants to work it out in out at the school gym armed with boxing gloves.

I was initiated the first month walking across campus to class when a seagull shit on my head.

Dick Cherry, a known tenth-grade agitator walked behind me, yelling, "wheelerbird, wheelerbird," while shoving me off balance.

Seeing the confrontation escalate to fisticuffs and attracting a crowd, Brother Paul the algebra teacher led us to the gym. Cherry had me by 20 pounds and had the reach of an orangutan.

Though my last boxing lessons by my cousin Jim's *Notre Dame* boxing champ roommate were three years prior, I still remembered his sage advice: "The target of the opponent's face is the nose."

I smiled.

As we sparred and parried, Cherry's overconfidence appeared. His movements were clumsily executed. Mine came smoothly and quickly. I dipped, arose, and steel-jabbed his nose, listening to the sound of delicious crunching cartilage. Blood spewed out his Cherry's nostrils and splattered on his face and T-shirt. He reacted with wild swings, and wrestled me to the floor, pelting my ribs until Brother Paul stopped the bleeding with a towel.

Cherry retreated in pain and ridicule.

I emerged triumphant, congratulated on my perfect shot.

Once I arrived home from school, I expected to see Ernie who was usually home waiting for his crew and get an earful about my fighting. The next home improvement project--renovating the living room to accommodate a family-size TV room and constructing an outside patio with a waterfall and pond. These involved knocking down walls, loading debris in the trailer, transporting it to a landfill, buying and loading wood and supplies, leveling the ground, and mixing and laying concrete in a black and white color checkerboard pattern, which Bill and I thought was nuts. We prayed for rain when the odds were against us.

However, Ernie wasn't there. I grinned and headed to the bathroom to wash my face off.

"It never rains in Southern California," lamented many workers.

"Never," another would reply, wishing for the rain that brings a day of rest.

Bill and I prayed that his dad wouldn't show up. We hoped Ernie would be delayed by staple calls, arriving home too late to force us to work. He sketched his plans on paper– designs that he pulled out when supervising us– but he never let us work unsupervised unless he had grunt work: moving dirt, rocks, lumber, or bags of cement. When it seemed clear that he wasn't coming, we retrieved our 22s, slung cartridge belts across our shoulders like banditos, and headed to the riverbed to plink and hunt jackrabbits, relishing the freedom to be children.

Postcard Legacy

The wise men brought gifts for Christ

on the day of his birth.

In 1954, they brought pain

I wantdered through weeds and debris,

stabbed trees,

nurled stones, fired at every living thing,

aiming with blurry eyes.

Aunt Marg said, "Do they fit, hon?"

Bill and I wanted hunting knives and ammo but settled for *Pendleton* shirts *and Levi's.* No card, gift, or phone call from Aunt Helen, Grandma Carney, my Kissing cousin Jane or dream girl, Carol. I wondered if Aunt Helen sent money to the Paskells for my keep? Marjean hugged an angora sweater and silver bracelet. Ernie was unwrapping a cheap ugly tie from Bill and me when the telephone rang.

Dressed in church clothes, blush and mascara in place, Aunt Marg pushed herself off the couch, walked to the end table, and retrieved the phone. The room was silent. She said in a cheerful tone, "Hello, Merry Christmas."

In the time-lapse of a photographer holding her breath, focusing on a subject, Aunt Margaret's glinting eyes shone a strange stare as if she was suddenly possessed. Her body stiffened, then trembled uncontrollably. Her mouth opened as if to yawn. But it was not a yawn of somnolence. What followed was the universal scream of death and overflowing grief. Aunt Marg collapsed to the floor. Ernie rushed to his wife and gently assisted her to the couch. I retrieved the phone off the floor and handed it to him. Marjean and Bill were frozen in silence.

Ernie said to the caller, "We'll leave right away."

I had not seen Grandma Carney in over a year--since the previous summer running free in Eastside Columbus. Christmas day was one week from my fifteenth birthday. I lived almost half of my life with Grandma Carney. But no one gave a flying fuck about my feelings. I found myself struggling for breath; my limbs turned to lead, preventing me from moving across the room to comfort my mom's reported "closest sister," but who, like other Carney siblings, never spoke of "Kathleen."

My early childhood days in the country flashed before me: Grandma Carney tucking me in bed, chasing away ghosts and nightmares, cleaning and bandaging my wounds, offering me the first slice of freshly baked bread, the first lick of chocolate icing, giving me a bear hug after "checking for "potatoes in your ears," and falling asleep while patching my jeans, fussing over me to stay out of harm's way. I was the center of her attention living at New Lexington, Grandview, and Eastside Columbus. Whenever I was with her, I never felt like a passed-around orphan nor was I treated like a black sheep, unlike at the Watsons.

However, I was old enough to understand Grandma Carney was elderly. She suffered growing health problems and was physically unable to function as a full-time caretaker for a "wild" kid with boundless energy, flunking school, and who was accident-prone. When she was too sad to listen to her favorite radio program, *The Romance of Helen Trent* or play checkers, she would commiserate over the death of Uncle Jim mining Carney coal, opening her leather coin purse and spilling the symbols of her grief: Army medals cherished more than gold Krugerrands. I understood when she called me "Jim," she was speaking to me as if I was her beloved firstborn son killed in the coal mine.

The night Aunt Marg and Ernie packed for Columbus, I sat alone in bed musing over a cigar box of personals brought from Ohio: *Sky King* ring, dozen Indian Buffalo nickels, a silver dollar from Dr. Spencer, a colleague of Dr. Watson who gave it to me after a painful penicillin injection in the butt with a spike-size needle while I was at home sick in bed. To this day, I am not sure which one– the silver dollar or the shot– sped up my recovery and return to school. Further down in the box, I found a black and white photo of me on a paint pony posing at grandma's Grandview duplex (I was seven) and a picture of *1951 CO-CHAMPIONS CATHOLIC ELEMENTARY LEAGUE LIGHTWEIGHT DIVISION ST. AGATHA SCHOOL* in which I am lost in the back row, jersey number obscured by eight-grade giants standing shoulder-to-shoulder in the tier below. Beside me was smiling fellow sixth-grade starter Timothy Donavon wearing jersey number 17. *Three years later*, I mused sadly, *I would become one of Timothy's pallbearers*. Even further down, I had tucked in a group photo of Cub Scouts *Den 11*: seven feisty boys in uniforms taken by our Den Mother, Aunt Helen. Finally, I found my most valuable memento: a color postcard of *Prize Palomino:* prick ears, posing in a fancy black leather western saddle with hooded stirrups, coiled lariat suspending from saddle horn, gold coat, white mane and tail meticulously groomed. He was a true premier stallion, with papers confirming his prestigious bloodline. I flipped it over to read the precious words scrawled on the back in sloping cursive.

Dear Jerry

Having a fine time. I am waiting on our dinner

So I thought of you and thought you would

Like this card and the picture of the pony

By

Grandma C.

A postcard postmarked *Albuquerque N. Mex Aug 12, 1953*-- The farthest west of Ohio that Grandma traveled, thanks to Uncle Joe's exploring a job offer weeks before his heart suddenly stopped pumping life. He was buried in New Lexington cemetery beside Grandpa Carney and Uncle Jim. Here

I was, two thousand miles away from Grandma Jessie Carney's funeral, who was more a mom to me than any of her damn daughters. I have no photograph of her, Grandpa Carney, Uncle Mutt, or my birth mother. Hell. I pounded my head on the pillow to sleep.

Next thing I knew, I was trapped in the smoke-filled cockpit of a flying fortress shaking from the turbulence of exploding Nazi flak. My slumped co-pilot's oxygen mask and wool-lined flight suit were covered in blood and shards of Plexiglas. I yelled "Bailout! Bailout!" in the radio headset over the cries and voices of a 9-man crew reporting doom. Wind from shattered Plexiglas and shrapnel punctures muffled my commands. Sudden yaw ejected me from the pilot seat and threw me against the kneeling, headless body of the turret gunner blocking the designated escape hatch. Instinctively, I dragged the parachute toward the navigator and bombardier compartment hatch in the nose of Flying Fortress. Blinded by smoke, fumbling with the chute harness, I felt my hands burning from a parachute pack engulfed in flames. My body levitated, slammed against a metal shell.

O screamed, "Bailout! Bailout Out!" I crawled out of the escape hatch, tumbling in the sky. Suddenly I jerked upward, arms and feet dangling like a puppet, body swaying downward, my goggled eyes madly searching for silk canopy, but only saw a burning wing of B-17 floating like a boomerang in slow motion, black buffs and winking stars. I spotted a human fireball in space falling toward me. The melting face was a dead ringer for me. Suddenly it was me, I was on fire.

I awoke to my screams. For the first time since I arrived in the Promised Land, the *Dream Catcher* hanging on the bedroom window curtain rod failed to catch a bad dream. Bill's folks left for Columbus in the *Buick* at first light. They arranged for Marjean to stay with the neighbor's family.

Bill and I relished an unfettered vacation since Ernie issued only one order: stay around the house and feed the family pets. Days of freedom awaited us. We hoisted up our GI field packs stuffed with *Saltine* crackers, *Spam*, *Skippy* peanut butter, *Butterfingers,* and *Baby Ruths*, flashlights and batteries, water-filled canteens clanging on cartridge belts bulging *Super X* .22 hollow point ammo, matches, and slung .22 rifles. Billy and I'd slipped out of our secret gate behind the house. We wasted ammo shooting rousted jackrabbits in furrows of the bean field. Jackrabbits proved elusive as crows. But killing soaring seagulls and sandpipers standing statuesque felt irreverent and disrespectful of nature.

Sitting on the concrete bank, we inhaled the salty air, feeling the wind awakening our senses. Looking skyward, we could track contrails of bombers jetting to the cold war. We'd sight our target, hold our breath, squeeze the trigger, watch a bottle explode, or can tumble backward from the impact of a hollow point-tip round, listen to zings of bullets ricocheting off the sand and smell the cordite. Sometimes we'd freeze in place, listening to keow of gulls changing course from inland to the sea, the hum of distant tractor-trailers delivering boundless merchandise and obsolescence to the edge of a new world. Bored, we'd shoot and miss crows foraging waves of debris that flowed in the Long Beach flood control ditch after storms. We explored culverts for treasure and bones, plinked some more, then gathered wood for a campfire. At night, we'd gaze at the stars.

The next morning, we hiked along the ditch like armed explorers. We eyed a building behind a razor-wire fence. A sign said: *No Trespassing US Government Property. Violators Will Be Prosecuted.* We found a crawl space. The faded gray wooden building with missing shingles and dirty windows must've been an abandoned secret WWII base. We busted the door padlock with a rock and entered the inner sanctum. Breathing stagnant air, we swiped away cobwebs at the entrance. The room had a high ceiling, our eyes locked on 40 mm, sawed-off brass shell casings loaded with pencils on top of dusty olive drab painted metal desks covered with airplane blueprints, and rulers. We scanned walls plastered with battle maps, photos of Nazi tanks, planes, V-2 rockets on launching pads and framed photographs of Presidents Roosevelt and Truman. We leaned our rifles against a desk, took swings of water from our canteens, gorged down Baby Ruth's, and scaled wooden steps to the second floor. I was the first to spot it.

"Dibs," I shouted and raced across the room to claim my prize. Bill followed and grabbed a pinup wall calendar of a stacked naked blonde wiggling her ass at intruders. We stuffed our packs with shell casings and the *1945* pinup calendar. With great difficulty and many built-in rest breaks, we also lugged a *Piper Cub* propeller home.

By week's end, we spent all our money on bus tokens to ride downtown, movie tickets, donuts, and *Orange Julius* and Pike Amusements Park rides. We plinked at the riverbed until we ran out of .22, .177 pellet gun ammo, resorting to shooting bottles with a Daisy BB gun and loading a pellet gun with a copper BB which was lethal for shooting sparrows, piercing tin cans, and destroying bottles. To be effective, an air rifle required 3-5 hard pumps per shot which I found was hard on the arm. Safety concerns stopped us from attempting to shoot flying birds and tossed bottles or cans with .22 rifles. Shooting BBs was another matter. BB gun fights-banning headshots- were common at the wooded banks of Scioto River. Stepping in the bean field, standing shoulder to shoulder. I was carrying a loaded pellet rifle, an air rifle, and a Daisy. Suddenly, Bill yelled, ***"Crow,"*** pointing his free hand at the crow flapping its gleaming black wings ten feet above. At the same time, I raised the Benjamin .177 cal air rifle and fired. ***Crack,*** loud as a lightning strike, then a human cry of, "You **shot** me!"

"Jesus! I screamed, dropping the rifle, staring at a BB-size hole oozing blood in the back of Billy's right hand. "Fuck, I'm sorry. I'm... sorry." Thankfully, the pellet gun was loaded with a copper BB, because I was out of lead pellets and loaded my rifle with a copper BB used for BB guns.

"Shit, my dad's gonna kill us."

"He gotta find out first. My best school buddy's mom is an ER nurse. Maybe she can help us. If not, we gotta see a doctor."

At home, I cleaned the entry hole with iodine, anxious that the BB was too deep to see. I expressed relief the air rifle was not loaded with a pellet that would have flattened out in flesh and risked lead poisoning. Luckily, that day I was out of pellet ammo.

An hour later, my friend's mom probed the wound, applied antibiotic cream, and covered it up with a band-aid. To our deep relief, she said in a confident tone, "This will keep your secret until it works its way out. Keep it clean and check on it every day. When you see a tiny gold bead, pop it out. And for god's sake, remember how it got there."

Our biggest fear was his dad confiscating our guns and condemning us to house arrest and a life of chores. Every day we examined the wound for signs of pus with a magnifying glass. Five days later, the evidence of my crime surfaced. We were in the clear.

Uncle Mutt Returns

Always a silver lining–

Death brings family together.

Bill's folks returned with Uncle Mutt who was left alone after Grandma's passing. Since he was Aunt Marg's closest brother by birth and knew Ernie who was raised in New Lexington before he enlisted in the Navy, it made sense for him to move in. I had just turned 15 and now stood eye to eye with my Uncle Mutt. Yet that didn't stop him from, at first sight, giving me the *knucklehead* treatment. I had a strong bond with Uncle Mutt that started in childhood. He moved into the TV room and slept on a sofa bed.

Not much had changed about Uncle Mutt. He still parted his slicked-back jet-black hair, gleaming with *Wildroot Cream Oil,* in the middle, and favored sleeveless undershirts, bib overalls, and high-top shoes. Now that I was learning my way around a razor, evidenced by several nicks around my shin, I noticed the way he swiftly shaved with a straight razor in a way that made me nervous. During his waking hours, he chain-smoked *Camels* with nicotine-stained fingers.

Still suffering from an unrepaired hernia, Uncle Mutt limped around the house and avoided heavy lifting of construction material used in Ernie's home improvement projects. He still helped out as Ernie ordered, and handed Ernie tools and hammered nails while Bill and I carried wood, rocks, and mixed cement after school. Aunt Marg continued working full-time at the bank while her younger brother spent his waking daytime hours in the TV room when Ernie was gone. His go-to programs consisted of wrestling and movies Addicted to smoking and daytime television, he made no effort at finding gainful employment. While money problems were never discussed by the family around children, I knew Uncle Walter had helped Grandma Carney with rent and bills when she was alive and made sure Uncle Mutt always had pocket money when odd jobs escaped him. I assumed his successful brother Walter continued helping his siblings in need, but I wasn't sure. In a short time, Aunt Marg had taken on two more. I hoped that Uncle Walter was helping out– Aunt Marg's life was already tough enough. She didn't need additional financial stress.

During my freshman year at St. Anthony High School, I boycotted school sports, tolerated classes, and carried out Ernie's work orders with Bill and Uncle Mutt. Luckily, a new classmate whose uncle was a longshoreman paid us each $100 a week after school for assembling a warehouse of English bikes packed in shipping boxes. Another buddy's dad who owned a Danish Bakery hired us to scrape clean baking pans on weekends. I was able to make enough money to reduce my dependence on the family's taut finances.

Life wasn't all school and work. During Easter break, we all squeezed in Ernie's *Ford* station wagon and towed a single bed rented camping trailer to the California High Sierra Mountains. Aunt Marg and Marjean slept inside; the guys roughed it in sleeping bags on the ground around a campfire tended by Mutt. Come daylight, Bill I armed with hunting knives and .22 rifles hiked the nearest trail, scanning tall pines for squirrels and crows nowhere to be seen. Weaving among the canopy of trees in the mountainous terrain of rocks and fallen logs, the rifles felt like weights on our shoulders and arms. Having experienced excursions in strange forests in Ohio, I always marked the trail with pieces of a bandana, conspicuous sticks, and stones to identify the return path. The elevation forced us to pause, catch a breath of thin air, and hydrate our thirsty bodies with water-filled canteens. Not a single critter or human was in sight.

After hiking another hour, we came upon a mountain runoff of freshwater cascading over smooth washed stone in the middle of pristine green forest. We splashed our lathered faces and drank pure cold water in cupped hands, gazing at small darting fish in the transparent stream. We removed our hiking boots and socks, rolled up our jeans, and waded to our knees. Soon, we were horse-playing, disturbing the other hikers who stood on the bank looking in awe of nature's lost paradise and tried to ignore our antics.

Suddenly leaning over the bank, pointing at the water, Marjean screamed, "I dropped my glasses." She was the only Paskell who wore prescription glasses. Responding to her cries, his complexion flushing crimson, Ernie rushed to Marjean's side and stared at her distraught face. Without her glasses, I saw from a distance her wide-open eyes shine with fear. We all knew what would come next, sensing the violence her simple error awoke. A loud smack echoed off the mountains. Ernie's arm catapulted into her head, striking hard and knocking her to her knees. Her tiny body was miraculously saved from falling into the stream by her mother scrambling to her side, sliding her arms under Marjean's sobbing body.

Ernie's hand pushed past her. He madly thrashed around the water as if the movement would force the glasses to the surface. Being located further downstream from the incident in waist-deep water saved Bill and me from escalating the violence to no benefit. We looked at one another in horror. Before we had a chance to exit the river, Aunt Marg quickly intervened.

Shoulders tensed, Aunt Marg examined Marjean's face and scalp while she sobbed. Then like many other previous episodes, she embraced her child and said in a soothing soft voice, "It's ok, Honey, just a dumb pair of glasses."

Marjean sobbed.

I stood there dumbfounded. *It was Aunt Marg's job to protect her child,* I angrily thought. *Grandma Carney taught me that when she prevented Uncle Mutt from whipping me. And I wasn't even her child. How could Aunt Marg allow that monster to strike her tiny innocent daughter?*

The family walked back to the trail, struck camp, and drove home in utter silence. Uncle Mutt occupied the front passenger seat, so Marjean could sit with her mother in the middle seat. Bill and I sat in the jump seats. On the way home, I pondered Earnie's patterns of physical violence. It was not exhibited toward his wife or others outside the family. At least I never witnessed him raise his voice at

Aunt Marg though I didn't know what words passed between them when they were alone. When he lost patience with Bill or me over chores, his fiery complexion gave warning blows were eminent. We braced ourselves or moved quickly from the area. Bill and I, being older and male, had the freedom to rush through chores and bike to the pier and bus downtown. We escaped his blows by limiting the time we spent at home.

Unfortunately, Marjean was younger and alone. Her body when her father was present, reminded me of a dog beaten by his master: flinching, hiding, and shriveling in his presence. She spoke cautiously, if at all, preferring to hide in her room or escape to her girlfriends' houses down the street.

The following week without explanation, Uncle Mutt took a Greyhound bus back to Ohio. Notwithstanding family ties, I figured the Paskells resented the extra work and cost of taking in a non-paying adult border addicted to watching TV's bellowing wrestlers' body slams, death locks, wheelbarrows, flying chairs, and tag teams crashes, canvas pins, fake screams, and takedowns submissions, and relentlessly used car commercials. The tension over Mutt's sudden departure coupled with Ernie's unpredictable depth charge explosions directed at the most vulnerable offspring was taking its toll on me.

Navy Contraband Ordinance

Paddleboards, umbrellas, and lounge chairs,

a perfect combination for horny sailors

and bikini-clad nymphs

Sometimes I'd be paid with money;

Sometimes with contraband:

liquor, cigarettes, and ordinance stolen from battleships & cruisers.

It was the summer of 1955, my second summer in California. I found a job serving tourists. Our dress code consisted of a white t-shirt, blue denim pants, and white *Converse* sneakers. Every morning at 9 a.m. I raced my Schwinn to a Belmont Shore workstation, unlocked a shed of 8-foot hollow wooden paddle boards, umbrellas, backrest props, and lounge chairs, dug holes in the sand with GI entrenching tools, and planted giant umbrellas with rainbow-colored petal flowers in arc rows. Next, I dragged and staked a lock box in the ground for customers' cash and drivers' licenses, and a logbook on rental time.

All day renters– horny sailors, bikini-clad nymphs, and adult couples with children– forked over money for time on the water or in the sand. I'd signal renters with a whistle that their time in the umbrella shade or paddling in the bay was up. When the sailors ran out of dough, they'd barter with contraband--ordinance stolen from Navy ships. A petty officer from Georgia gave me a sock of live rounds of 50 caliber ammo to rent a board for a blonde all afternoon. A gunner's mate bartered for an umbrella and props with 20 mm and 40 mm shell casings. The brass gleamed like the copper tone bodies of girls as they lay on boards paddling nowhere, while lobster-skin sailors made ripples close behind them, chasing their fantasies.

After I told sailors that I lived with my ex-WWII submariner uncle, they offered to give me a tour of a cruiser. They drove me to the Long Beach Naval Station and put me on a ship-to-shore skiff to their ship. I was beside myself scaling a plank aboard, strictly against Navy regulations. They bribed an ensign whistling approval for a civilian kid to board with a giant bottle of *Chivas Regal.* Sailors let me peek inside an 8-inch gun turret they said was fired on *Iwo Jima,* let me operate an elevation crank of a double-barrel 40mm rapid-fire anti-aircraft gun that shot down *kamikazes* in the *Battle of Okinawa.* They took me inside the *restricted* radar room that scanned the sea and sky for invading Russians. The sailor in charge sat at a green iridescent radar screen that flashed blips, lines, and intersecting symbols. But what impressed me the most was the loaded 45 semi-automatic pistol on the counter within reach of his hand and the 24-hour armed guard outside the hatch entrance.

At sunset, I'd strike the set: remove each board's drain plug cork, stand it on end and drain pounds of water, collapse umbrellas, fold back prop and lounge chairs then store them in the shed. I'd count the day's take and wait for the owner to arrive. On a good day (weekend, influx of sailors on shore leave) the money added up to $200 or $300. The red-headed crew cut owner was a local high school football coach. He paid $20 cash for a *good day* and $10 for slow. I maintained a leger on rental items' time of start and return and kept a driver's license or military ID as a deposit.

One could skim by not logging in customers. But this was never a temptation for me. The legal working age in California was 16. I was not 16 yet, but I was paid in cash. We both benefited from the arrangement.

The best youth jobs after this one were serving as box boys at *Safeway*, and packing up food for middle-aged women. At this job, I earned a fair wage while getting to stare at scantily clothed teenage girls. Besides, my fellow workers had warned me that our boss spied on us to assure we did not grant favors to pretty girls or friends or left our posts. We peed in the shed or had a buddy guard the lockbox when we used the *Union 76* across the street to take a dump.

On my third day, I spotted the owner standing in the back of his pickup parked across the bay, scanning the rental station with binoculars. I vowed to follow his rules.

Once the owner picked up the cash– usually by 7 pm– and paid me for the day. I'd hop on my bike, stop at *Joe's corner drive-in*, gorge on salty stringy fries and greasy burgers and drown in ice-cold root beer. Then, I'd wash away the grease and my sins in the ocean bay, race home, listening to brass jingling in my backpack and puffing away at a cigarillo given to me by a sailor for another free hour of shade. And like he was, I dreamt of golden-haired nymphs with big tits and apple-shaped asses, slim legs curled on beach towels under blue, green, and red canvas flowers or straddling paddleboards in a riptide, waving and screaming to be rescued, all the way home.

Of course, a side benefit of this job was my improved physique. Biking to work, and lugging heavy paddleboards in sand hardened my arms, back, and thigh muscles and improved my endurance.

My younger cousin Bill was not as fortunate in earning pocket money; earning a pittance conscripted by his dad's new early morning venture – delivering *The Los Angeles Times.* My summer job exempted me from losing sleep to toss L.A. printed news out of car windows of a fast-moving *Buick* driven by a madman through the maze of track homes. He made far less money for more difficult work.

During summer days off and school day weekends I continued playing touch football with buddies at Bellflower Park. Here I met a friend Donny Albert who attended Wilson High school. His parents owned and operated Danish Bakery on Long Beach Blvd. They emigrated from Germany weeks before Hitler declared war against humanity and freedom. Donny said his grandfather who lived with them in Germany and America shook hands with Hitler at a 1939 Nazi rally. The irony was Donnie's dad joined America's Merchant Marines and was torpedoed twice by Hitler's Nazi Wolf packs by delivering war supplies to the Allies. The family bakery was famous for its bear claws and wedding cakes. Donnie and I were paid 1 dollar an hour for scraping oven-baked sugar off cooking trays. The Alberts proudly drove a new Lincoln. On Sunday, Mr. Albert paid Donnie, who was 16 years old, and me each $5 to drive the Lincoln to the drive-through car wash. Donnie's folks awarded him a new red

1955 Corvette convertible at Christmas. After that, he regularly gave me a ride home after work. We made plans to meet up with girls at the drive-in burger joint, but Ernie encroached on my limited leisure time. He forced my participation in his never-ending home improvement projects. Sadly, I turned Donnie down.

My California exile stunted my growth- (5'7 in.,140 lbs.) and intellectual development. My new school required more rote learning and regurgitation of Latin, World History, and typing. I spend many boring hours calculating algebraic equations, solving geometry problems, and listening to classical music instead of Chuck Berry, Nat King Cole, Four Lads, or Sinatra. Impatiently, I waited for the 3:30 p.m. token-ride escape.

Ernie taught me to drive the stick shift *Ford* station wagon soon after my sixteenth birthday, though he didn't teach me out of the kindness of his heart. I got my driver's license the same day he ordered me to take over the paper route the next morning. He woke me up to greet the darkness of the morning and handed me the key to the *Buick*. "Bill will show you the route," he said and walked away.

We popped the bundled wire of two stacks of *The Los Angeles Times*, folded each paper, and then piled them in the back seat. Bill sat in the back seat and rolled down the rear windows. I turned on the headlights, gunned the engine, and put the automatic Dino flow transmission into drive.

"Turn right," Bill shouted.

We entered a labyrinth of narrow streets paved as an afterthought. Narrow streets were bordered by undistinguishable track ranch-style houses; many without street signs or house numbers. I felt the wind brushing my face and heard the swish of flying news out the rear windows.

Bill said, "Turn left next fucking corner." He must've had a photographic memory of the route; papers flying from both of his hands. "Jerry, turn left at the corner. Go faster."

Gripping the steering wheel, I scanned the road for parked vehicles, construction debris, feral cats, and dogs, weaving in and out of strange roads without street lights, guided by the *Buick's* bright headlights and my cousin's staccato instructions punctuated with profanity and laughter. It was still dark when we rolled into our driveway. Bill boasted over the record speed of covering the paper route. We didn't notice the *Ford* was gone until Ernie stomped in the kitchen door a few minutes behind us. He stopped in front of me, his face turning from light tan to beet red, his fiery eyes burning a hole in mine. He thrust his right hand, poking my chest. He said in an aggressive tone, "Give it to me."

"What?

"Your driver's license," he snapped,

My brows furrowing in puzzlement, I looked at him from the corner of my eyes, my silence igniting a short fuse of a live bomb.

"Drivin' like a goddamn maniac. So fast I lost you."

I retrieved my wallet and gave him the license. There was no point in arguing.

I failed the first driving test. But I passed the second time. I said to myself, *the son of a bitch tried following us the whole fucking route to check on us.* But the upside of Ernie's punishment was never again awakened at 4 fucking A.M. to deliver newspapers.

That spring we heard *Millikan,* a new public high school, was opening in the fall in the neighborhood and was recruiting football players. Every day at St. Anthony's, I encountered cocky jocks wearing purple and white leather-sleeve letterman jackets, strutting into class and across campus like celebrities. St. Anthony's had a history of Catholic League football championships. And their top athletes were on track to earn college football scholarships. Feeling like an outsider, I boycotted St. Anthony varsity games. At the same time, *Bellflower Park* tag football rekindled the flame of organized football and reminded me of my St. Agatha's glory days. I responded to an invitation to meet prospective *Millikan* high school football players and coaches who encouraged us to get in shape by joining our current school's track teams, then contacted one of *Millikan's* assistant coaches to work out in the summer before reporting to *Millikan's* August practices.

Joe Peters was St. Anthony's head football and track coach, and one of the few lay teachers. He was short, had dark beady eyes, and military crew cut and barked history battle dates and orders like a movie drill sergeant. Students called him Mr. Peters in class and Coach on the practice field, and "Mister Prick" behind his back because of his proclivity for the testing threshold of human pain at practice and was a hardass grader in *World History*. He drove the school bus after school to the Varsity football stadium and track field across town. I planned to divide my time evenly between the bakery job and track practice as much as possible to appease Ernie's obsession with us boys earning money. By working Saturdays and one or two weekdays cleaning bakery pans kept the Paskells in bear claws. After track practice, I hitchhiked home or took the bus, whichever arrived first.

Ready to get started, I bought a pair of spike track shoes and jogged loops until sidelined by side aches. All in all, I was a flatfooted sophomore reduced to a 10-yd. sprinter, crash-prone hurler, clumsy pole-vaulter, and pathetic broad jumper. My salvation was a hulk St. Anthony varsity senior football tackle shot-putt mentor called "Ox." I managed to medal in the junior varsity shot put competition. Ox picked me up in his chopped-body 36 cherry-red *Ford*, as I was hitching home from practice. Its chrome grille gleamed like a whale's teeth. No door handles. Custom pearl *Naugahyde* interior. Angora's dice-winking snake eyes dangled from the rearview mirror, and a floor shift plastic knob had a picture of a naked woman. Ox stopped at Wilson High and picked up a cheerleader with long golden hair who hopped in between us.

A few minutes later he careened around a corner, and suddenly braked 50 yards from the top of the steepest grade street in California, and grinned, "Ever flown in a hot rod, Wheeler?"

Before I could respond, much less comprehending his meaning, Ox revved the engine, then popped the clutch in first gear, leaving smoke from his tires after peeling out. We riders screamed simultaneously, catapulting into the sky. Our bodies levitated in mid-air. I swallowed my intestines. Our skulls slammed against the headboard. Wavy hair blinded me temporarily. Her shoulders and legs smashed into mine as we landed hard on the car's plastic cushions when we reached the bottom of "Airplane Hill." When Ox moved his hand from the gearshift to the cheerleader's chest, I stole the plastic knob of the woman with vibrating tits, escaped out the door, and ran home.

The next day, I bought bus tokens for the rest of the year.

The Cherry Theatre

Every Wednesday and Saturday I arrived early,

put in my time doing whatever needed to be done,

advertising films,

selling tickets,

serving refreshments,

escorting elderly ladies to their seats

while they remarked, "You're such a nice boy."

Once relieved by the manager,

I could live out my fantasies with the stars.

Chugged a Cherry Coke.

Making the varsity cut in August required taking advantage of summer workout sessions offered by Nick Bright, Millikan's future assistant coach. He scheduled morning drills, running plays, passing patterns, and ten-yard sprints, ending in two-mile runs at Long Beach State College practice fields on lower campus two miles from the Paskells. This meant I had to give up my lucrative day paddleboard job and find a night job. Luck waved her hand at me from a classified ad in the *Long Beach Independent*. I found work as a night movie usher and concession worker position at the *Cheery Theater* in the Long Beach retirement neighborhood. The hours were 4 P.M. to midnight seven days a week; 85 cents an hour, furnished uniform: black sport coat, white shirt, clip-on black bow tie, casual slacks-no jeans, and hand-held flashlight. In contrast to the discount movie house catering to drunken sailors and Marines with hookers lured across the street from the Pike Amusement Park, the *Cherry Theater* catered to mostly senior citizens residing in the community associated with the historical *Townsend Movement* that advocated old age pensions before FDR passed Social Security. Every Friday, I strapped on a sack of the alphabet, climbed a ladder, retrieved a letter, and spelled a fantasy on a marquee. Afterward, I dressed in a white shirt, black, clip-on bow tie, and a hand-me-down tux, sat in a glass booth, and sold quarter tickets to mostly old folk from the neighborhood. I raced behind a concession stand, juggled *Ju Ju Beans, Good &Plenty, Milk Duds, Boston Baked Beans*, shoveled popcorn in bags, and made the change for the pop machine. Then, I grabbed a flashlight and led late arrivals in the dark to a seat in the proper theater.

Sometimes I had to break up necking teenagers or call an ambulance for nursing home residents. On breaks, I danced with Kim Novak, kissed Ava Gardner, marched down the aisle with Grace Kelley, and rode a carousel with Shirley Jones. During the bus ride home, I gulped a cherry coke and waited anxiously to flirt and hold hands with Marilyn Monroe the next day (three times) after I changed the marquee for a matinee.

On the fourth of July, Bill and I drilled a hole in a six-inch length of half-inch pipe, loaded it with a firecracker, pulled the firecracker fuse through the drilled hole, tightened a nipple breech, added .22 gunpowder packed with cotton, and loaded barrel with a ball bearing. We mounted the barren on a block of wood, aimed the miniature cannon at the redwood fence, and lit the fuse with a kitchen match. Two planks of redwood fence disappeared in the smoke of cannon blast. The recoil landed us in the neighbor's shrubbery, where Ernie found us hunkered down, busting a gut before he fired the next round: "You idiots! Replace missing planks. Then weed dichondra and wash and wax the cars," I was forbidden to drive because, in his eyes, I was an unsafe driver.

The next day, I was too beat after morning football drills to carry out Ernie's orders to dig up old sprinkler pipes before going to work at *Cherry Theater*. I looked forward to my last *Picnic* dance with Kim Novak then changing the marquee to the *Searchers* and *Rear Window*. That night I fell asleep on the bus and crawled into bed. Suddenly I awoke to the bellowing voice of the ex-submariner meting out punishment to his insubordinate ward.

"Young man, consider yourself grounded. From now on, the only ground and morning drills for you are digging up old sprinkler pipes and installing a new sprinkler system. *Millikan* High School summer football drills are not in your future."

His words were as sharp and lethal as an executioner's swishing ax decapitating the condemned. But the executioner had no black mask hiding his blood-red face, snarling expression, and fiery eyes burning with sadistic pleasure I had seen before. I remained silent throughout, but I vowed to myself that he would never hurt me again.

In a TV nature documentary I saw during the previous month, fledgling eagles start branch-walking to build up their wing muscles and confidence before soaring out on their own. I figured two fucking years of slavery was damn long enough. One hour after Ernie left the house to make his *Bostitch* stapler rounds, I retrieved a *Sunoco* map and marked the *Route 66* escape route with a grease pencil. I outlined my plan to his son Bill because it would fail without his cooperation. My three-year age advantage allowed him more freedom to venture with me to the Pike, fishing pier, plink, and roam the neighborhood without adult supervision. Bill understood that his dad would track me down the minute he discovered I was on the run and may resort to contacting the police. Thus, I needed to have a minimum 24-hour head start.

"We'll dig up a few sprinkler pipes today to appease your dad enough not to wake me up just to chew my ass before he leaves for work tomorrow. Today, I am packing my stuff, cashing my last check, paying a coworker gas money to drive to Upland where I'll start hitchhiking East on *Route 66*."

I instructed Bill to convince his dad and mom I was sleeping in when they headed to work. If all went according to plan, at night, they would assume I was selling popcorn and candy and helping old folks into their seats at the *Cherry Theater*. By that time, if lucky, I'd have made it to Amarillo or St. Louis and called my Aunt Helen.

"What you gonna say?"

"Your dad's fucking crazy. If he ever hit me like he does Marjean, it'd surely come to blows, then he'd put me in reform school. I figure Aunt Helen will let me at least stay with her until I am 17 in January and sign my Army enlistment papers. I am sorry I can't say goodbye to your mom and Marjean."

Sagging his shoulders, Bill said, "My dad will kill me, thinking I covered for you."

"Bill, just act surprised. I didn't tell you shit. Hell, he may not figure it out until he contacts my boss at Cherry Theater who'll just say I am AWOL from work."

The Great Escape

The runaway hitched a ride outside Bakersfield

a hot, dusty town

where all residents were temporary

Handsome driver in overalls

with jet-black hair and dark eyes.

Took in a kid on his way to Ohio,

ready to join the army,

anything to escape an Uncle who

couldn't raise a dog.

My life and future were stuffed in a duffle bag: a commando knife, football shoes, one pair of *Levi's,* three-day changes in socks, shorts, T-shirt, jersey, GI field jacket, poncho, stocking cap, a disassembled *Winchester* model 69 rifle, two loaded 10-shot clips, *Weaver* scope, water-filled canteen, dozen saltine restaurants packs, four apples, *Butterfinger*, licorice sticks, cigar box of personal mementos, toothbrush, toothpaste, dish towel, and hand soap.

I sat on my duffle bag in the desert heat outside Needles thumbing on *Route 66* after eight hours of rocky rides with strangers sharing their life stories of surviving on the edge of the earth. One AWOL GI asked for gas and beer money before he dropped me off outside Upland. Another roustabout whose wife left him for a wildcatter, cried for 25 miles while he spilled his pain. A tattooed ex-con driving a jalopy accompanied by a bulldog he rescued, offered me cigarettes and beer before the engine threw a rod in San Bernardino and I was abandoned on the shoulder. Finally, a blonde woman wearing a blue bandanna dressed in a loose halter and shorts driving a yellow *Cadillac*-top down-convertible to her estranged father's funeral.

She called me, "Hon," shelled out $20, kissed me on the cheek, and said, "I hope your Aunt takes you back."

My biggest fear was spotting a red *Buick* that I washed and waxed but rarely drove in the rearview mirror. However, all I saw were outlines of tall saguaros guarding a flat landscape, a black sky winking tiny stars at a distant mountain horizon. As evening lazed into the night, I pondered my two options: standing on the shoulder risking becoming roadkill, or sleeping on the desert floor off to the side. I lifted the duffle bag, ready to surrender to the night's creeping headlights and the drone of 18-wheelers speeding east without me. The sound of a backfiring pickup rolling to a stop 10 yards away made me rethink those options. At first, I was not sure if the driver was making an emergency stop or just teasing me and would take off once I approached. He flashed his lights and I hustled forward.

The pickup was a vintage heavy-duty *Dodge* dually. It had dented fenders, tattered mud flaps, wooden side rails on the bed, hauling junk, and scrap iron. The passenger seat cushion was covered by a folded army blanket to protect the rider from being poked in the ass by coiled wire. The vibrating engine was shaking cigarette butts out of the beanbag ashtray sitting precariously on the dead dashboard. The cab was dark; smelled of oil and cigarette smoke. My night vision revealed a weather-faced young man with bushy eyebrows, and slicked-back hair, dressed in work clothes.

"Where ya goin' kid."

"Ohio."

"Lowell," he said, extending a greasy hand.

"Jerry, I said, exchanging steel handshakes as Uncle Mutt taught me.

"Git ya to *Grants,"* he said, the truck's engine sputtering like a beating heart in cardiac arrest.

Scanning the lonely highway through a cracked windshield, he grabbed the knob of trembling floor shift and winced each time he double-clutched and pumped the accelerator, surging his human and metal cargo forward on *Route 66*.

I explained my plan to join the army if my aunt did not want me back and signed the papers when I turned 17. As vehicles sped past the glare of passing headlights revealed a driver with gleaming perfect teeth. When we stopped for chow at a roadside saloon east of Gallop, I observed Lowell favoring his left leg as he exited the truck and dragged his foot across the parking lot crowded with pickups towing horse trailers, and 18-wheelers packed with cattle and merchandise. Finding no vacant tables, we sat on barstools beside truckers and cowboys. I ordered a hamburger, Coke, and fries; Lowell asked for ham, eggs, and black coffee. He sat next to a broad-shouldered man overdue for a shave, wearing a grimy straw western hat, faded work shirt, dirty jeans, and dusty western boots.

Out of the blue, Lowell poked the head-taller driver with his elbow and said, "Bet I can endure more pain than you."

"Bullshit," the cowboy snapped, retrieving his wallet, and slapping a five-dollar bill on the counter.

Lowell, stone-faced, whipped out a pocket knife, opened the blade, raised the knife, then without a blink, stabbed himself in the leg. Ignoring the wide-eye cowboy, Lowell removed the knife, slapped his thigh, picked up $5,

"Cork leg," he grinned. "Kid you're lookin' at the original three-minute warrior. That's how long a fuckin Marine machine-gunner lasted at Iwo. So..o, whatever you do, Kid, don't ever volunteer to carry a fuckin **B--A--R**." Probably a vet himself, the cowboy didn't complain or ask for his money back.

Lowell dropped me off east of Grants as the night rolled in like a black tarp over *Route 66* dips, curves, rangeland, mesas, and ancient tribal communities frozen in time. The only star in the thundering sky was a *Texaco Star* at the last gas station. Foolishly, I lost the gamble against catching a ride before pelting rain, a falling desert night temperature, and fierce winds that stopped highway traffic and drove me off-road, squatting and shivering under a GI poncho, field jacket, thin jersey, stocking cap, and sneakers. I felt the water seep into my sneakers and jeans exposed by the flapping poncho. Rocking to stay warm, I faded in and out of consciousness, awakened hours later by my coughs, painful leg cramps, and the sound of whining tires of 18-wheelers making waves on the highway in the pooled water. Standing on the shoulder in the headlight glare of passing semis, I shook the water off the poncho, coughed up phlegm, and spat it in the sand. The sky was still pitch-black. I retrieved a flashlight from the duffle bag and shined it on my watch.

"Shit," I screamed into the night sky. Here I was at 4 A.M. standing on the gravel shoulder next to a duffle bag, facing oncoming cars. I looked young, but I had no choice but to extend my arm out to communicate with benevolent passerbys, thumb-jabbing fate.

Fate turned out to be a compassionate *Peterbilt* fuel truck driver delivering a tank of gas at *SUNOCO* stations in *Albuquerque*. The cabin was warm and the seat comfortable. The dashboard displayed countless luminous dials and a two-way radio. I fell asleep in a steel womb; awoke at the last refueling station. Concerned over feeling woozy waiting out the storm, lost hours, and increased risk of crazy Ernie catching my runaway ass, I thanked the driver, found a phone booth, and told the operator I wanted to make a collect call to Hudson 8 5224 7744, Columbus, Ohio. "Tell whoever answers, the caller is Jerry Wheeler."

"Yes, operator. Honey, Are you Ok? Where are you?"

I coughed, "I am ok. Hitched rides to Albuquerque."

"Been waiting for your call. Do you have money?"

"Yes."

"Jerry, listen to Auntie Helen," she said in a supportive-authoritative tone. "Go directly to the *Greyhound* bus station. Buy a ticket to Columbus and call me the minute you arrive at the bus station. I'll pick you up. Any problems, call me immediately. Understand?

"Yes."

Army, Nuts!

Maybe Aunt Elizabeth will take me back,

Maybe she'll sign the papers.

My future hangs in the balance.

Please.

Please,

anything but send me back to Ernie's

I awoke to the sound of swishing doors and the bus driver's voice telling Columbus-bound passengers to retrieve their luggage. I had no memory of the ride from Albuquerque to Columbus Greyhound station. I sipped an ice-cold *Coke,* feeling a curative effect on my dry throat, called Aunt Helen, and waited outside scanning the street for the sight of a 1954 metallic green *Buick Special* with gleaming whale tooth grille, white sidewall tires, and *Dynaflow* transmission she drove to daily mass.

A few minutes later, she slid up to the curb. The car was immaculate as new. Suddenly unsure of myself, I wondered, *should I appeal to her, asking her not to hand me over to her psycho brother-in-law and let me stay until January?* I took a deep breath, grabbed the shoulder strap of the duffle bag, opened the backdoor, and tossed it inside. She gestured for me to sit in the front seat. I quickly obliged, observing her welling eyes as she scooted across the front seat, and embraced me.

She said, "Honey, I am so, so sorry for sending you there. Ernie Paskell is not fit to raise a dog, let alone, children."

I nodded and looked at her more closely. The only evidence of change in her appearance was a few strains of gray hair. She wore her public face, spotless, pale blue summer dress, and her presence gave off a perfume fragrance that hopefully masked my rumpled sweaty road clothes. She dabbed her eyes with a tissue, as we sat in silence. Grandma Carney, having passed the first Christmas of my exile, her firstborn daughter, Helen, held my future in her hands.

She could have returned me to face Ernie's reprisals. Though we both knew, I wouldn't stay long at the Paskell's house. I would either escape again or be sent to reform school. Would she let me stay if I promised I'd be a delayed volunteer ward of Uncle Sam? She had kicked me out before. Would she be willing to let me come back?

After a deep breath, I looked into Aunt Helen's brown eyes, and said, "Aunt Helen, can I stay here until January, when you can sign my Army papers?"

"ARMY... ARMY...., are you NUTS!" She gasped as if reacting to an incredulous proposition. Then she grasped my arm tightly like a mother stopping a toddler from crossing a busy street, alone.

Jesus, I said to myself, expelling deep relief from fear of Ernie's wrath, but dreading more years of compulsory all-boys Catholic school education at Columbus *St. Charles* or *Xavier,* Yet, knowing that *Upper Arlington Senior High* was good enough for her son Jim who went on to graduate *Notre Dame,* entered the *Dominican* seminary, and fulfilled his mother's passionate dream of her son's priesthood gave me hope. Maybe she would let me attend a public school if I attended mass with her. Her cloistered daughter Jane was away at a Jesuit College looking forward to future coupling with "a good Catholic boy," unable to accompany her mother to weekly church events.

No Japanese manufactured model plastic American warplanes suspended by a thread from the third-floor bedroom ceiling, model ships and vintage cars parked on the dresser or ink-stained desk, or tart board, and a *Hormel Chili* miniature college pennant collection pinned on the bathroom door. The attic room looked as sparse as the day I moved in. I set about unpacking my few belongings, making the space homier.

At 6:30 P.M., Aunt Helen opened the kitchen door, peeked at the cheek of her husband returning from the hospital, and said in a casual tone, "Honey, Jerry's back."

Three inches taller than the passed-around Carney clan orphan grandchild-in-law, the quiet doctor and surgeon with a patient list of Ohio's movers and shakers and the first osteopath physician appointed by the governor to the traditional exclusive MD Ohio State Medical Board, extended a firm hand. We shook hands as if I had returned from another summer sojourn at Grandma Carney's Eastside duplex. The husband-wife ritual at the kitchen door was soon followed by my devout Catholic aunt reciting grace and our subsequent silent dinner.

In terms of physical appearance, the passage of time only revealed more salt than pepper in the doctor's conservative hairstyle. He sported the same visible left side part, length cut above the collar, bushy eyebrows, beady eyes diagnosing thoughts and flaws behind the thick bifocal lens, wireframe clinging to the patrician nose, complemented by a modest mustache, a Wednesday golf-day tan face, and stature projecting confident authority of an English Prime Minister. As in past summers, he was impeccably dressed in a seersucker suit, Chinese laundered pale-blue *oxford* cloth button-down shirt, bow tie, and polished *Florsheim* classic wing backs that he traded for soft *Keds* before retreating into the den.

Grueling Two-A-Days

My high school teammates called me "Wheels"

as I sashed through Tigers & Bears on Friday nights,

suddenly braked, backpedaled, bolted forward,

zigzagged past tacklers to TDs

like a wild man on a unicycle.

Two-a-day hot August football practices were held at Northam Park, a mile from the new Senior High School already called "The Factory" because of its two-story red brick rectangular architectural style. The official football stadium remained at the old high school converted into a junior high.

The road to self-validation following exile and escape began at the Golden Bear football team's physical wet weight recorded as 164 lbs., 5'11inches. Many aspiring teammates knew me from my former grade school days playing touch football at Northam Park, or from my spectacular cannonballs at *Scioto Country Club* or *Arlington's Municipal* pools. Locker room snapping towels and antics felt like a reunion. But, in contrast to St. Agatha's days, making the varsity cut as a *Golden Bear* was far from assured.

Upper Arlington Senior High was a serious contender in the *Central Buckeye League*. The Golden Bears football team had three varsity coaches. The head coach, Marvin Moorhead, was a tall, blue-eyed, square-jawed, hard-bodied man with a blond buzz cut that gleamed like a cue ball in the sun. A face tan as a farmer who spends days driving a tractor, baling hay, or mending fences, he appeared to be in his late thirties. He spoke with a gruff voice whether praising or chewing out an individual player. Most days, he sported a *Property of Upper Arlington High School* athletic jersey, shiny silver whistle dangling on a chain from bull neck, knee-high khaki color football pants sans thigh and knee pads, and high-top *Spaulding* aluminum-cleated football shoes. According to rumor, he was an ex-college lineman, committed to Woody Hayes's, "Two yards and cloud of dust philosophy." Mostly, he served as the linesmen's coach and defensive coordinator.

Coach Moorhead had two reactions to completed plays. After a perfectly executed play, he'd pull the involved player by his shoulder pads, shouting, "all right." If the play was bungled, he'd grab the culprit(s) by the shoulder pads and shove him away in disgrace, grunting, "no, no, no."

After practice, players stood at attention while he delivered a critique, flexing his pectorals and making his points. Assistant coaches, Hank Cummings and Pete Kerry favored school jerseys, shorts or baseball trousers, and low-cut cleated shoes. Hank was short, soft-spoken, and had a dark flattop haircut. He was the team's trainer and running back coach. Pete, a stocky, towhead blond, wore shorts, an athletic jersey, and a black *Golden Bears* baseball cap. He was the quarterback, kicking coach, and varsity baseball coach.

Intent on seeing "Wheeler" written on Moorhead's office chalkboard listing the first-string players' position on his two-platoon system, I brought back the sage advice once given to me by a St. Agatha veteran: "Impress the coach in head-ons." Hunched in the middle linebacker position, wearing a defensive team jersey, I locked my eyes on the squatting quarterback's eyes while he scanned opposing forces. Chewing a rubber mouthpiece, I waited for the quarterback to shout words, "hut, hut," signaling the center's snap to the quarterback and the latter's handoff to the fullback bolting through a narrow passage before being slammed shut by a human wall of padded flesh. The ball carrier running on instinct was oblivious that he would encounter a teammate's exploding rage and anger. Rage and anger suppressed over years of passed-around existence were ready to burst out of me. I was searching for my place in the world; the guilt of never asking my kin who Robert Vernon Wheeler, the name on my birth certificate, was. I recently found it among Aunt Helen's papers. Was he a scoundrel, army deserter, pedophile, or habitual criminal? I was furious with myself that I hadn't pursued this.

All that anger accumulated in my tense muscles as I crouched like a wild predator, feet planted on turf. I pounced on my prey, driving a plastic helmet and shoulder pads into the runner's chest, wrapping tenacious arms around his legs, lifting, and holding his body as if offering a human sacrifice to the gods of violence. The Gods listened to the clash of flesh, bones, and plastic armor and voice roaring catharsis of the orphan's unconscious turmoil. The roar could no longer be muffled like I had smothered my childhood nightmare screams by my pillow.

I had learned my trade at Columbus Eastside's *Roosevel*t *Recreation Park* and a four-year grueling apprenticeship at *St. Agatha's:* a tackler never deceived by the runner's head dips, and turns, shifty eyes, shoulder, and arms feints. Only leg movements telegraph the path.

The next play brought an adrenaline rush. I leveled the Golden Bear's senior star's pass receiver, my helmet hitting him in the spine, his body carried off the field, never to return. Though my wish was to feel the pigskin tucked in my arms, it was temporarily denied.

"Fuck**....**, let me off, here," I ordered Pat Patton, a teammate driving me at the end of my first two-a-day practice when I saw Ernie's *Buick* parked in the Watson's driveway. I hid behind a neighbor's tree until the *Buick* disappeared. "Olly, Olly, Oxen Free, mother fucker," I mumbled at his retreating car.

I entered the house. Aunt Helen made me a sandwich and sat down at the table with me. She said Ernie wanted her to turn me over to him, but she told him, "Ernie, you had your chance."

I thanked her and counted myself lucky. Bill and Marjean were stuck. I was able to escape, and my aunt was willing to hide me. She didn't tell me she loved me or seemed happy I was back. But she didn't want to send me back to the evil monster.

Aunt Helen wasn't perfect; I always wondered why Grandma Carney never visited Watson's home in contrast to Dr. Watson's aged and feeble mother who came over several days a week. Also,

Aunt Helen never attended the annual Carney clan reunion at Somerset, as if she was ashamed of her country roots. But I was appreciative, nevertheless.

Two-a-day drills, running laps, sprints, blocking dummies, and contact scrimmages were welcome outlets for venting pent-up rage over involuntary exile, grief over Grandma Carney's death, and attending a Holy Cross Brothers Boys high school I hated. I was always a serious athlete and my main cheerleader. Within the Carney clan and peers, there was no stigma attached to being an orphan or notion that my institutionalized mother and Aunt Wanda, my cousin Anne's or Uncle Mutt's respective afflictions connoted shame or fear of contagion or less human value. I was aware early of the consequences of *death in the family* before I could tie my boots or read. I observed the grief in Grandma Carney's eyes the first time she called me "Jim" and showed me her dead son Jim's Army shooting medals. I had met Grandma's brother, Jim Webb, a nearly blind mustard gas survivor of *The Great War*, influenced by the name of Grandma Carney's first-born son, my Uncle Jim, and perhaps was instrumental in Aunt Helen christening her son, Jim. I wasn't Jim, but I came from a line of tough, successful fighters. A fighter I was through and through. I was willing to fight for my place on the team.

I had known Pat Patton, UA's varsity backup halfback since Scioto Country Club swimming pool days. The first time I visited his home, I was taken aback by three framed photographs on the stairway wall: a young fighter pilot climbing into the cockpit of a *P-40 Warhawk Flying Tiger* renowned for its shark-faced nose art; the identical pilot sitting in the open cockpit of the same plane flying in the sky; a group photo of *Flying Tigers* pilots posed in their distinguished leather flight jackets. Suddenly, the pilot in the photographs descended the stairs. Star-struck, I belatedly shook the ex-combat fighter pilot's hand. Both Pat and his dad were tall blondes. I felt shame at the fantasy lie I told GIs on the train that my missing dad was a bomber pilot killed by Nazis when a real-life hero stood in front of me. Mr. Patton did not have or need a regular job. This allowed him to attend our practices. His freedom was understandable to any Ohio meat shopper. Mrs. Patton was the daughter of *David Davies Meat Packing Company's* owner. She was extremely gracious, welcomed Pat's friends to overnight stays, and played pool in a knotty pine-wall well-stocked basement recreation room. Pat drove a *1955 Chevrolet* station wagon with a bull horn as he rousted cattle to the slaughter at his summer job.

After several two-a-days, on Day 4 we tiredly completed our morning ritual of entering the locker room and then checking the names posted on Coach Moorehead's office chalkboard.

"God Damn," I said.

"Fuckin Aa, Go-git-em Wheeler made it," Pat said, jabbing my arm.

Waiting for the next play action at the afternoon scrimmage, I shifted my linebacker's eyes from the breaking huddle to the blonde standing statuesque at the sideline, her ponytail swishing in a gust of August wind. She was dressed in white tennis shorts and a matching sleeveless blouse.

"Shit," I said, thinking, no way she'd recognize the face of her former soulmate in bulky pads and headgear.

The runner, starting right halfback and fellow junior, Phil Patterson, the son of the Ohio State diving legend and founder *Patterson Pools,* shot through the line. No malice intended, I hit him laterally in the secondary so hard we landed on the sidelines. We removed our mouthpieces and helmets, shook off the lime dust then trotted past the bystanders. I was about to raise my hand to attract Carol's Sievers's attention when she acknowledged Patterson with a puppy love smile, as he dashed to the gathering huddle.

"Absence makes the heart grow fonder," was bullshit. I had no cause for jealousy, having made no effort to contact Carol Sievers for over two years. Faces and memories fade over time. But, during a break in practice, helmet in hand, I could not resist approaching her. We warmly embraced each other. She said she was happy that I made it back safely; my Aunt Helen called her at home on my second day on the road, asking if I had contacted her. Carol heard I was back from a girlfriend who dated a teammate.

Seeing my name posted on the defensive lineup the last days of August, I was haunted by the memory of waiting two years to break into the backfield at St. Agatha's. Moorhead came across as a rigid coach. Once he determined a player's best fit in his two-platoon system brain, the peg stayed in the same hole. Indeed, the "Gogitem Wheeler" moniker worked against crossing over to offense. The final 30-man roster names appeared carved in stone. My fullback debut was a result of the starting fullback sustaining a debilitating leg injury during a scrimmage in which I gave him no quarter. Pat Patton's dad hounded backfield coach Cummings to give me a chance to fill in.

The play call was "ten-right on `hut` three." I assumed a three-point stance, and waited for "Hut, Hut, Hut," then catapulted left off the tackle behind pulling left guard leading the way. Clutching the pigskin, I bolted in tight space between the center and left guard leading interference.

Entering the secondary, I tucked the ball in my left shoulder, grabbed the guard's rear hip pads, and shoved him into pursuing linebacker tumbling backward. Suddenly, the pursuing safety materialized and dismissed my feigning side-step.

Instinctively, I raised my free arm, and swung a fisted hand like a medieval war weapon, hitting the side of the defender's plastic helmet, knocking him on his ass, his headgear flying off, rolling in the end zone like a foe's severed head. I relished my victory; I was an essential member of the team. My mates clapped me on the back or slapped me on the ass, saying, "Wheeler, damn. Just you out there."

However, we acted very differently when walking through the formal halls of school. The Upper Arlington Senior High School's dress code banned jeans, T-shirts, and sneakers. We all wore plain color Ivy League trousers sporting decorative belt buckles on the back that never made any sense. Who likes to lean against a chair with pieces of metal pressing into their ass? Shirts were flannel, colorful cotton prints, oxford clothes, and V-neck sweaters. Dr. Watson's custom-made shirts were my "reserve." Shoe options included dirty bucks, saddle shoes, and cordovan brogues purchased at *Gilbert's Discount*

Shoes. All sports footwear was bought at *Grandview Outdoor Store*. The entire ensemble was quite different from our Friday Night Lights uniforms.

My high school classes were not as tortuous as classes had been in elementary and middle school. Heretofore, a decade of rigid Catholic education offered limited options in the curriculum. What was offered was of little interest to me. The stern nuns and Holy Cross brothers rewarded conformity and rote learning, neither of which I particularly liked. Pity the poor bastard who could not diagram a sentence or compute a math problem at the blackboard at chess-clock speed. Public shaming ensued for losers instead of Holy Cards. Recess and Catholic theme movies like The *Story of Saint Bernadette*, and *Boys Town* were welcome diversions. Students craved holy days called "free days," and snow days that closed schools.

Public schools provided a head-spinning number of options. UAHS offered two official tracks of curriculum tied to academic test scores and grades: *college prep* and *"other."* One look at the battery test results taken at St. Agatha and St. Anthony grades convinced UAHS guidance counselor Lance Schreffler that the college prep track: advanced algebra, trigonometry, and physics, would place me in harm's way. *Passing to play* sports was implicit. I was good at regurgitating memorized facts, poor at math, and lost in geometry and chemistry was impossible. But failing a course was not an option. Thus, I needed a fail-safe formula. Mandatory core courses were English, Civics, American History, Chemistry, and Physical ed. Shop or vocational education succumbed to working-class prejudice. Other elective courses had me in mind: Speech & Drama, Art, Choir, and Journalism with school newspaper assignments.

After two years of passive involvement and self-imposed anonymity in the Holy Cross Brothers educational process, my new life and freedom at the Watson's rekindled my passion to play organized football and explore new subjects.

I felt validated after hearing the UASH football stadium's public address system blare, "Ball, carried by Wheeler.*"* In 1956, before face bars were standard, I started the first six games as a fullback until my nose was smashed in a head-on collision with a Mount Vernon two-hundred-pound linebacker and waylaid in a double-team tackle that broke three ribs. My proudest day was walking the *Factory's* halls wearing a Golden Bear black letterman's jacket with a golden ***A*** patch. Like the social pecking order at St. Anthony's, jocks ruled. Football was the most prestigious sport at Upper Arlington Senior High. But the biggest sports star and senior was Jack Nicklaus whose hands were saved for golf and basketball, not pigskin. But golfers were not viewed as heroes by the girls. The rumor was that Jack's father, who owned a chain of drug stores, did not permit him to play football because of the risk of a hand injury. His golf teacher, Jack Grout, a pro at the Scioto Country Club gave lessons to the club's youth members, members that briefly included me before I was a caddy who admittedly had no talent or temperament for the sport of gentlemen.

Honky-Tonk Encounter

Strangers be damned,

we were from the wrong side of the tracks.

Would we be sentenced to our deaths or

a little bruising?

Staying to find out: not an option.

With our knuckles bleeding, my teammates Joseph Decker, Sam Ross, and I dragged our cleat-gashed chins and torn cartilage earned from too damn many head-on tackles and pile-ons into a honky-tonk bar to celebrate Golden Bear's Friday night triumph. We wore our letterman's jackets and ordered 3.2 beers with forged draft cards and smoked cigars.

Suddenly a bearded guy with greasy ducktails walked up to us and growled, "What's the fuckin' `A' on the jacket stand for... Asshole?"

"Look like a bunch of assholes to me," said his buddy in a *John Deere* cap, dirty overalls, and hayseed breath. They eyed our buzz cuts, clean Levi's, dirty bucks, must've figured correctly we were snobby Upper Arlington rich kids on the wrong side of the Scioto River. I signaled our school's short safety, flashing my eyes to be "cool," and slowly retreating from the bar. With every step we took in retreat, the agitators grew and multiplied.

"Let's get these pussies!" A stranger's voice shouted.

The second we hit the parking lot, we dashed to Decker's WWII surplus *Willie* Jeep he and his dad assembled in the summer. He leaped behind the steering wheel and nervously started the engine. I followed, landing in the other front seat, deflecting flying bottles with a raised arm. The sounds of exploding glass bottles, jeep's grinding gears, burning knobby tires spitting gravel, vehicles revving vengeance, backfiring mufflers, and screams from hell thwarting our escape.

"Fuck! Fuck! I said loudly, while Joseph sped across the Scioto River bridge while the strangers chased us. I was hurling socket wrenches, pliers, and a hammer retrieved from a toolbox behind the seat at the pursuing headlights and windshields; Joseph was losing a car race on Scioto River Rd. Suddenly, Decker jumped the curb, driving 4-wheel drive off-road, diving into dips, flying over rolling hills, weaving among trees in the rough, knobby tires spitting dirt and fairway divots, fording streams and mutilating greens of the 9th and 10th holes of the exclusive *Scioto Country Club's* golf course.

I flipped the middle finger at a pickup scuttled in the pond behind the pro shop, and a metallic blue *Mercury* spinning its rear wheels in a sand trap. Ross waved at a customized Ford coup that crashed into an oak at a dogleg curve, smoke rising from the broken radiator and crunched hood.

Bouncing in our seats, we laughed until our bellies hurt more than our football wounds and waved at a customized Ford coup that crashed into an oak at a dogleg curve.

Once we were safely home, we three swore off violating training rules for the rest of the gridiron season.

Self-Prescribed Academic Formula

I chased down and pounced on runners, busted noses

with forearm blows, clawed gouges in legs,

knocked out teeth

as parents shouted from the sidelines.

Comforted their children after the game.

No one ever watched me.

My academic failsafe formula included Civics, sports reporter of the school's newspaper- *The Arlingtonian-,* a thespian-cast member of *Time Out for Ginger,* and the worse voice of 175 members in the prestigious traveling *Upper Arlington Senior Concert Choir* modeled after Fred Waring group that traveled around the state, including sharing the spotlight with Governor C. William O'Neill addressing 12,000 teachers at St. John's Arena. The choir was led by Herbert Yenser, a handsome, dark hair, pale face man in his late thirties whose only handicaps were a high-pitch voice and a withered left arm that hung down his side like a scare crow's straw-filled coat sleeve swinging in sync with the movement of good arm directing the choir. The only reason I passed the audition, I was a popular jock. The choir's favorite annual excursion was to Huntington W. Virginia. After a performance standing on risers before an audience of Mason conventioneers was stopped by a prankster's exploding smoke bomb, we retreated to motel rooms, changed from black choir robes to waiter's white aprons made from sheets. Then we returned to the ballroom, gathered leftover liquor and snacks, called our female counterparts sworn to secrecy, and threw a wild party. Other pranks followed: students smuggled Tommy-gun squirt guns in violin cases into choir practice and carried out St Valentine's massacre. Wearing a mask and black cape, dressed as Zorro wielding a sword, I chased the class magician and his fancy costumed female assistant across the stage at the students' annual talent show. The raucous audience thought it was part of the act. I finished the year as a graceless dancer in Yenser's spring musical production, *Carousel.*

I got an Ohio driver's license to run errands but depended on friends' transportation. My first date was initiated by a Golden Bear senior cheerleader with an Olympiad body that cartwheeled over hurdles. She picked me up at my house in her dad's new silver *Buick Lesabre* and drove us to the *Scioto River* where we watched the submarine races. She had a detectable mustache in the moonlight and kissed me until it hurt. I was obsessed with football more than girls until the season's end. Taking booze or smoking cigars pilfered from the surgeon's den was off-limits. A wealthy all-white school district, UAHS sponsored a plethora of intellectual, creative, and physical activities to lure students across the physical, creative, and intellectual bell curve. Upon reflection diverse educational programs that were conducive to spoiled tumultuous adolescents' successful transition to responsible adulthood. I observed potential valedictorians in the homeroom packing their resumes for acceptance to prestigious colleges, sororities, and fraternities. These goals or objectives, I had no more interest in learning than the names of minerals listed on a chemistry lab poster. UAHS offered over half a dozen scholastic groups ranging from the *National Honor Society* to *Quill and Scroll,* Foreign language clubs in *French, Spanish,* and *Latin*. It had a *Debate Team, Science, Literary, Drama , Library, Future Teachers of America, Future Nurses of America, Charm, Red Cross*, *Camera*, and a *Radio Club.* Their *Festo* Club oversaw Christmas decorations and *Y-Teens*. The all-boys *Hi-Y Club's* motto was a fucking joke: *clean living, clean speech, clean sportsmanship, and clean scholarship.* The lame administration's newest approved girls club was called *J.U.G.* Boys countered with *The Boner Club.* Emulating the ostentatious OSU Marching Band, the Golden Bear band opened every home football game led by a tall baton-wielding drum major strutting across the field in Friday night lights.

Students gravitated to tribes. Football stars occupied the top tier on the social mountain. Honor society members were viewed as "goodie-goodies." Theater Arts attracted athletes and scholars alike. One could count the outliers on one hand: guys that wore western shirts, jeans, and engineer boots to school dances, avoided sports, worked on old cars in school parking spaces, and kept to themselves. My social network spawned from football teammates, fellow thespians, and buddies that joined the choir as a lark or for the girls.

One of my best friends, the football team's manager, golf team member, and classmate was Trent Lamb. Far from an impish stereotype image of a "water boy" catering to the equipment needs of self-absorbed gladiators, Trent possessed the unique skill of shoring up players depressed over girl troubles, a game loss, and chewed out by Morehead. Trent filled the supporting role of a brother, employing humor and optimism. He was a great organizer and recognized as a vital member of the UAHS championship golf team. He called me "Wheels" on the gridiron and "Buffoon" for my growing reputation as a prankster and class clown.

Having transferred from Westside, Columbus, Trent was one of the few classmates exposed to working-class people. His dad was an Ohio public servant; his mom started her advertising agency. They lived in a modest house in the expanding Upper Arlington community and welcomed me as a frequent weekend guest, a pattern that developed among new high school friends. Aunt Helen had no qualms over me staying overnight with friends. But reciprocity was out of the question.

Luckily, I made friends easily, allowing me to cycle through their houses on weekends and never overstaying my welcome. Football sidekicks included Pat Patton, a starting defensive linebacker. Bob Neal was a reserve lineman whose strongly independent mom was building a secret log cabin house herself in the woods. Varsity halfback star, George "Crazylegs" Collins weaved and hurdled tacklers like L A Rams' "Crazylegs Hirsch." Visually handicapped, George inserted oversize contact lenses in his eye sockets before a game. A fastidious procedure in front of a mirror his teammates found repulsive but enabled him to see where he was going. George's younger sister, Susan was as beautiful as their blonde stay-at-home mom who welcomed guests following Friday night games. His dad was a prominent physician who gave George a used *Hudson Hornet* we called the "purple whale."

I had some pocket money, money that enabled me to join my friends for entertainment. My earnings consisted of a $5.00 a week standard allowance, plus a $2 monthly haircut allowance, an extra *Lincoln* for cleaning the garage or shoveling snow away from the driveway, driving stake holes with a sledgehammer to fertilize the trees, and spreading potting soil.

My aunt drove me to school on her way to daily mass, but I was dependent on friends or my feet at dismissal because there was a social stigma in high school against riding bikes. Students without personal cars for school were picked up by their parents. Aunt Helen was an outlier among adult women of her social station. She shunned martini luncheons at the *Club,* and golf matches, and never attended a movie or sports event other than Notre Dame when her son Jim was a student there. Her leisure time was monopolized by church activities, advancing Doctors Osteopathic Hospital, and managing household needs. She strongly believed the catholic religion provided the best moral code for adults and a yardstick for raising children. She also stressed the importance of Catholic education. She was openly critical of Upper Arlington parents that "materially spoil their children." She encouraged self-reliance in her kids and her ward. During summer college breaks, her son Jim was a busboy at the local Italian restaurant and worked at a Fruit company in Hawaii with his Notre Dame roommate. Jane had a stint in her dad's hot basement hospital kitchen. Consequently, unlike most of our peers, Trent and I took a job at *Italy's Deli* during the 1956 Christmas break and were expected to find summer work.

Wearing white shirts, paper hats hiding our burrs, paid 65 cents an hour, we'd hunch over the freezer, cold hands grasping curved trowels. We'd stab ice barrels hard as concrete filled with chocolate, cherry vanilla, napoleon, peach, and lemon-flavored ice cream. We'd scoop skyscrapers in waffle cones, hoping our creation didn't tumble from the tiny hands of fidgeting children, their snowshoes dripping slush on the shiny linoleum floor we mopped an hour before. I'd slice skin-thin, and weigh fifty cents worth of baloney, pimento, roast beef, and Swiss cheese for impatient customers. I prayed I didn't cut my fingers off while Trent shoveled mountains of coleslaw, potato, and macaroni salad into leaky cartons with wire handles. We poured rivers of steaming cocoa and coffee into Styrofoam cups; punched keys of the cash register next to the empty tip jar and counted change with a forced smile. Last crowded Saturday afternoon, half of the Golden Bear football roster and their girlfriends charged into the deli and ordered, hot fudge sundaes with whipped cream, nuts, and cherry on top" from the fullback and team manager, I handed the quarterback the scoop, Trent lobed cherry, chocolate, mint, and vanilla ice cream grenades at teammates, cheerleaders, and the homecoming queen dashing out the door.

Until I met Brenda Clooney, a free-spirited, spoiled brunette sophomore at the monthly school dance, no one had caught my eye since Carol. Brenda lived in a mansion on Lane Avenue, living the lifestyle of a Hollywood star. According to co-eds, she had a collection of cashmere sweaters that assured she'd never wear the same one twice. She often drove an *Oldsmobile 98* to school but insisted that I drive it on dates and to her home after school where we cozied up in a basement entertainment room: bar, TV, telephone, pool table, and most importantly, jukebox containing the current rock and roll and slow dancing 45-rpm hits. Her mom was tall, attractive, flamboyant, and trusting of her daughter's freedom and our hibernating smooching. Her dad was friendly, letting me drive the *98 Olds* home and return it the next day.

He offered me a summer job working in a concrete crew of his construction company which was a better alternative to scooping fucking ice cream at minimum wage or caddying duffers' for tips. The aloof Watsons were never curious why an *Olds 98* was parked in their driveway on weekends during school or a 1952 cement-spattered *Chevy* pickup truck bed loaded with float tools, a power smoother, spreader, come-along-tool, water buckets, shovels, and a sledgehammer was parked in front of a luxurious colonial home.

Because of him, this was the summer I got in shape for those late August two-a-days and my senior year gridiron season. On hot days, I rose at 6 A.M. and laid concrete slabs on basement floors and driveways. Measuring and laying yards of drain tile and spreading concrete in underground rooms was the hardest task because the basements were used as latrines by construction crews and the dank air made you puke.

The white concrete foreman was an ex-WWII *Screaming Eagle* paratrooper that survived the *Battle of the Bulge* without a scratch. He said the most professional army was the Wehrmacht, but the most fearsome soldiers he'd never want to face, the Russians. He liked to work in a hurry. He said, "wading in cement was better than wading in mud, blood, and shit."

I was the youngest of three white laborers and the only one that had a driver's license. We picked up two middle-aged white guys that lived in a rooming house near the OSU campus. They reminded me of the hard Appalachian faces of convicts I'd seen at the pen in earlier years. One guy rode in the bed of the truck until the foreman went on a three-week vacation. He put me in charge of the truck, equipment, calculating and ordering drain tile and cement yards, and weekly payroll. We broke our backs, scooping and shoveling lead to keep from being buried alive when concrete trucks' chutes poured cement thru basement windows. Then we'd hunch over, smooth out bumps with two-by-fours, stretched muscles we didn't know we had, and sweat as if in a sauna. For a finishing touch, as the temporary crew boss, a powerful machine with propeller-size blades whirled me around the room until I was too dizzy to stand and nearly died of exhaustion. Driveways weren't much better with all the digging, making forms, and hoping it didn't rain in the middle of a job.

Only on these nights, I never dreamt of sex with a woman or awoke with a boner. When football practice started, I was in the best goddamn physical shape in my fucking life. I had high hopes of leading the Golden Bears to a championship season.

Late Hits

Her eyes, gleaming like neon.

She had cameo skin in a pink strapless formal,

ruby lips, a wrist blooming roses

and breasts the color of bubble gum

swelling with attention.

Drawing attention from all the men.

She was all mine–

only for a night.

I was the unchallenged fullback starter until a sophomore reserve with the same surname, vying for varsity promotion drove his shoulder pads into my left knee after the whistle-ending play during a scrimmage. The lethal late hit tore the cartilage as my body was relaxed. Maybe this was not the tackler's intent, or I was too cocky and deserved the hit. Regardless, I was caught off guard. I temporarily hung up my cleats, limped, and pickled my body in daily whirlpool treatments for two weeks. Game nights, taped ankle to mid-thigh, slowed me down. As soon as I returned, I was sidelined again with "Crazy Legs" Collins and several varsity players by unrelenting flu which sadly cost us three games.

My worst personal defeat was against St. Charles 33-35. It was the only football game my Aunt Helen attended in my seven-year football career including games with St. Agatha's. Motivated by her best friend Peggy Bettendorf, mother of Bobby Bettendorf, a best grade school friend, and St. Agatha teammate and St. Charles's halfback star, she willingly dragged herself into the stands. Arlington was on St. Charles's three-yard line with five seconds remaining on the game clock.

Our quarterback, Bob Jenks, whispered, "Now is the time to execute the secret play- the 'fullback pitch.'"

The "fullback pitch" was a play we practiced every week but never called in a real game. The defense anticipated a fullback plunge or "Crazy Legs" Collins end run. The secret play called for a fake handoff to left halfback Collins around the right end. My assignment: set up a yard behind the quarterback, step forward, pause and slowly fade left, and quickly turn to receive the pitch behind the left end as the halfbacks and line movement shift right. We stood in the huddle, ready to execute the play when Coach Morehead in his usual fucking hardhead fashion, suddenly sent in a substitute player to override Jenks's call, devouring the vital seconds needed to snap the winning play. After the game, Bettendorf and I communed in the end zone, reflecting on the many seasons we donned the same scarlet and gold uniform and shot each other in BB gun wars at the banks of the Scioto River.

During my senior year Christmas vacation, I scooped and floated concrete driveways after thawing frozen ground with a campfire, pickaxing rocks, and leveling dirt to pay for the headlights and bumper of Trent's mom's car I wrecked making a 3.2 beer run during the Thanksgiving break. Meanwhile, I stopped dating the *Cashmere Queen* to pursue a new girl who appeared on the scene on the last free day of summer of 57.

Gazing out the third-floor window, my eyes zoomed in on a slender, long hair teenage girl in tight jeans and a blouse. She was unloading boxes from a station wagon and carrying them inside a three-story red-brick mansion constructed behind the Watson's property during my California exile. On the vacant lot, I dug foxholes and bunkers to play war as a kid. It faced Lane Avenue and Scioto Country Club's stone border fence that Jane and I often scaled with our sleds after snowstorms. Her name was Bobby Edmonson. She was a junior from Southern California who moved here with her divorced mother taken in by Bobby's maternal grand folks.

Fortunately for me, the only accredited course that included students of lower grades was the *Upper Arlington Senior Concert Choir.* Bobby was assigned the first row. At the sound of the bell ending the class, I rushed down five risers to catch her walking in the hall with a group of attractive junior class equals that included cheerleaders and the girl's swimming team's champion diver. I introduced myself as her neighbor who observed her unloading the car the day she arrived in Upper Arlington.

The following Saturday, I walked across our connecting backyards and knocked at the front door that faced Lane Avenue. A tall, attractive woman in her forties, an older version of her tall and slim youthful daughter opened the door. She was wearing a *USC* sweat jersey, pressed Levi's, and flip-flops. She scanned my flat-top, dark eyes, raised brows, letterman's jacket, and smiled. She led me arm-to-arm into the plush living room where Bobby was sitting on a sofa, perusing fashion magazines. She looked up and quickly shoved a stack of mags away. She motioned for me to sit beside her so close our jean-covered thighs touched, and I smelled the scent of perfume. We talked about California: how she missed the beach and friends but felt very welcome here.

"I have no car," I told her, "but I would meet you at the gym before Friday night's home football game."

"That's sweet," she said. "But Gary Silver has already asked me. He's picking me up."

I grimaced, feeling the sting of rejection. "That's ok. Another time." Gary was a junior who lived the next block over. We continued small talk until I made an excuse to leave.

Trent dropped me home at midnight after the basketball game and a hamburger at the local *Frisch's Big Boy* at Northwest Blvd. Never subjected to a curfew by my guardians, I removed my shoes and tiptoed in the dark to my room. Still reeling over Bobby's rejection, I paused at the top of the third-floor stairs and peered out the window into darkness when I saw a car's headlights beams entering her driveway. The car stopped adjacent to the attached 2-car garage that had a night light above the door. Her house was dark. I speculated that the car's occupants were Bobby and Silver. My heart racing, I saw a tall figure emerge from the passenger side in the glare of the night light. She entered the house through a separate door by the garage. Shiver maneuvered his car in the double-size driveway to make its exit on Lane Avenue. I hoped Bobby would not turn on the house lights obliterating her night vision. My imagination and testosterone propelled me to leave my window perch, retrieve a flashlight and return to the window. Focusing on outlines of a large residential structure and distant trees in the moonlight a thousand yards away, I aimed the beam at the second-story windows of her house and started signaling in the dark like a sailor adrift in a life raft at sea. Suddenly, a tiny light flashed several times from an upper-floor window. I answered back, flashing three times, holding my breath until three blinks answered. I directed the beam from her window to Bobby's backyard and held it until our beams intersected.

Within minutes, we played flashlight tag, running across the frozen grass, and slipping between shrubberies bordering our property. We stopped to catch our breath at the three-foot brink terrace wall dividing Watson's backyard from the others. We removed and folded our coats into seat cushions and sat shoulder-to-shoulder, dangling our feet. Our eyes intently studied each other's faces; we listened to silence and our heavy breathing, hands attacking buttons and zippers, then fingertips feeling soft golden skin revealed in the moonlight.

Notwithstanding our passionate beginning, Bobby rarely turned down dates from other students. Strong-willed, her feminine independence rejected the notion of ever going steady. But she rarely missed our midnight rendezvouses. On sub-freezing nights we'd race to my garage and lay in the back seat of the Doctor's 1955 *Buick Roadmaster*. But, to our chagrin, the car's interior was too freezing cold to shed winter coats. Jealousy toward my male competitors led me to steal and plant the city's giant metal ***Dip*** street sign warning *road hazards*, in a prominent rival's front yard. It was a two-man job. The main culprit was no secret. Most rivals were not deterred. But the big prize was accompanying Bobby to the Junior/Senior prom after a formal dinner at the Scioto Country Club. It was the only occasion that Aunt Helen offered me her car to use for a social function.

1957-1959

The Visit

Feeling abandoned and alone,

as the dark day, I was mysteriously taken away.

It rained so hard, mixing with my tears.

While I thought occasionally about my biological mother during the previous decade, I experienced minimal remorse regarding my biological mother's fate. No childhood emotional bond developed between the two of us in the first five years of my life. Physically moving from New Lexington to Logansport was a lost memory, buried so deep it was inaccessible. My actual memory of her starts with suddenly living in a strange house with a bizarre acting woman– my birth mother, and her husband, a kind railroad man named Earl.

I felt more of the painful psychological effects of two abrupt separations from Grandma Carney, grief on her passing, loss of contact and affection of Carol Sievers, never getting to see my close Cousin Jane, and the shock of rejection from Aunt Helen sending me to live with an abusive ex-submariner than I did from not seeing my mom. Adding to my anguish of being uprooted from my close Upper Arlington friends and teammates, was being forced to attend a regimented Catholic Boys' School.

All I wanted was to belong, and I tried my best to make Aunt Helen proud, so she would keep me. I studied hard and was passing school. I was an athlete, a four–year football letterman eager for the next athletic challenge. She didn't have to do much for me. Hell, I fixed my breakfast, lived with friends on weekends, never asked her for rides outside of school, and earned pocket money caddying and odd jobs.

When she, the Carney matriarch, cast me out of her home, sending me to a tumultuous home, I felt the true sting of rejection. Being "tired" seemed like an excuse. It was one of the worst indignities of my childhood. Jesus Christ! All Mighty.

Notwithstanding, though I was treated as the Watson's black sheep's ward, I deeply understood the material advantages of living in a white privileged family. Aunt Helen didn't indoctrinate me with Catholicism as she did with her son Jim. She smothered him, placing him on the path to the Dominican priesthood from the day he was born and continuing as he served as a St. Agatha altar boy, and finishing as he graduated from Notre Dame University, and entered the Dominican Province. She treated her daughter Jane similarly. Every bit of Jane's life was tightly controlled. Her education started with a day-school nunnery- *St. Mary of the Springs-,* continued with a Catholic high school, and finished with a Catholic college until she met and married a "good Catholic boy."

I rarely thought of Kathleen, but my biological father was not even on my radar until the kind GI on that Columbus-bound troop train made me realize I was a six-year-old kid alone among young adult men, men my supposed father's age. Heretofore, I was accompanied by loving kin in the New Lexington homestead surrounded by father figures such as Uncle Mutt, doting maternal Grandparents, and Uncle Walter. After ten years of no contact, Kathleen Carney Boyer was a faded image of a distraught, half-dressed woman stumbling around the house, eyes blinking like a bird that crashed into a glass window; she was bewildered more than cognizant of reality or aware of her bizarre behavior. At best, during her breakdowns, I was a bungling child first responder, anxious for stepdad Earl or his folks to intervene. At worst, left the house until she calmed down.

To my surprise, the week after Easter in my senior year, Kathleen came back into my life. I was eighteen. Aunt Helen summoned me to the den and said in an adult tone, "Jerry, you have reached the age to take over some of Auntie Helen's responsibilities helping your mother."

She relayed the years visiting my mother at the *Ohio State Mental Institution* on West Broad Street, sharing how she apprised her of my progress through photographs taken at Grandma Carneys, St. Agatha's, and Watson's. She described how Uncle Walter helped their youngest institutionalized sister– my Aunt Wanda, Uncle Joe Carney's widow Evelyne and my cousin Anne, and made sure Mutt had a roof over his head. They both deposited money in Kathleen's and Wanda's commissary accounts to purchase cosmetics, cigarettes, and female niceties not provided to mentally ill wards of the State. This was one of the rare times that Aunt Helen handed me her car keys.

I walked to the car unsure of how I felt. The drive to the mental institution to see Kathleen- my biological mother passed by in a blur.

Soon, I pulled up to the massive ancient stone building on West Broad Street that reminded me of the Ohio Pen minus the walls and bar-covered windows. I went to the visitors' desk and asked to see the social worker for Kathleen Boyer. A middle-aged woman dressed like a schoolteacher wearing a name tag, *Joan Whitman MSW,* pinned to her blouse introduced herself as the social worker in charge of the ward assigned to the oldest residents, located on the fourth floor. She said, if I had called ahead, they would have kept Kathleen back from her job. Consequently, the social worker said my mother would meet me at the bench in front of the laundry building where she worked.

A gust of cold breeze brushed my face. I paused outside the main administration and ward entrance. Gazing at the blue sky and scanning a sprawling landscape of century-old trees guarding buildings gray as father time, I checked my jacket's pocket for the pack of *Pall Malls* and a book of matches, given to me by my aunt.

Before I left, she handed me both. I looked at her puzzled. "Patients are not allowed to carry matches, so you'll need these; plus, your mother goes through a carton of cigarettes I leave at the commissary in a week. After they are gone, it's *Bull Durham."]*

I followed a footpath as directed for ten minutes, leading to a large brick building with protruding pipes and rows of windows. I approached a figure of a crouched woman rocking on a weathered wooden bench, both hands gripping the bottom slats as if holding on after tired of standing or afraid of slipping off the seat. She acknowledged my presence, catapulting from the seat, touching my face like a sculptress working clay, and running her hands over my football arms and shoulders. Her dark brown eyes welled up behind thick black plastic state-issued glasses.

She cried, "Oh, hun, you look so growd up."

Her hair was coiled gray and white wires partly hidden by a plaid scarf. Her creased face was smeared in makeup nervously applied on the fly to meet her son she hadn't seen in over a decade. Smudged red lipstick clashed with nicotine-stained chipped teeth and fingers of addiction and neglect. She grasped my hand and pulled me to the bench. We sat side-by-side; I was still while Kathleen repeatedly crossed and uncrossed her legs, shifting her glazed medicated eyes downward from my face to the floor. Her state-issued black-laced work shoe pumped in and out of the sun's glare. Her opened coat revealed a faded denim dress rumpled by steam and heavy work at the laundry asylum.

She said in an embarrassed tone, "They didn't tell me you were comin'."

I apologized for the surprise visit, averting my eyes from her twitchy face. Late, I learned later that was tortured by years slaving in a snake-pit, her body injected with enough thyroxine to put down a Clydesdale, and her brain subjected to lobotomy that failed to mend a broken mind. Before I offered a *Pall Mall* as Aunt Helen instructed, my mother uncrossed her legs, sat up erect, and retrieved a drawstring pouch of *Bull Durham* from her coat pocket. And, with the dexterity of a Texas drover, she untied the pouch of tobacco with fingers chewed to the quick, removed a single cigarette paper from the pouch, and placed the drawstring between her teeth, letting it dangle below her chin as she a formed a crease in the paper. Next, she grasped and re-opened the pouch, sprinkled a line of tobacco in the crease as if it was gold dust, closed the pouch with her teeth, and returned it to the coat pocket. Then she smoothly licked the edges of the paper, rolled it as I'd seen in western movies, and inserted it between her thin lips.

She flickered her eyes at sky and me and said, "Gotta match? We aren't allowed to carry them."

I stood facing her, lit a match, and concentrated on holding it at the hunched figure leaning into the flame. She took a long drag, smoke drifting out of her nostrils and cracked mouth, enveloping her face and stooped shoulders. The smoke smelled strong as burning leaves, causing my eyes to sting. I sat beside her again and engaged in small talk I do not recall.

A fast-burning fuse racing between her nicotine-stained fingers, my mother said, "I sort ladies' things." When the rollie's dying ash touched her fingertips, she squinted at the blue sky and resumed pumping her legs.

My body vibrated on loose planks from the movement. Unlike pipe tobacco or cigar smoke, which I found pleasurable breathing in the doctor's den, my lungs strained not to explode from the cigarette's toxic fumes. Reacting to her frantic movements, I stood, and retrieved the pack of *Pall Malls*, tapping the bottom until a single unfiltered mass-produced cigarette emerged.

Freezing her legs, she held a *Pall Mall* to her lips, leaned into the flame, and inhaled deeply, her eyes flickering with pleasure. When the last ash reached her thumb and index finger, she dropped the tiny butt and resumed pumping her legs, her empty hand gesturing for another cancer stick; switching off the current to her legs in the first drag. This ritual: pumping legs between smokes, went on until her sad eyes suddenly clicked shut, and she slowly rose from the bench, stared at me as if I were a stranger, turned aside, then charged the fortress-like-building with protruding steam pipes exhaling puffs of smoke like cannons engage in friendly fire. I watched my estranged mother disappear into the haze of insanity and smoke. Thereafter, scheduled visits started with me ascending a dark, winding stairway like the *Tower of London* to her ward's waiting room, then leading the descent to the same park bench where I engaged in small talk and fed her addiction.

To me personally, these visits were not an emotional reunion over traumatic parent/child separation. My brain censored out or redacted saying the word mother aloud because of life experience. Kathleen Boyer was a member of the Carney clan, not to be forgotten, disparaged, or treated with unkindness over afflictions beyond her control any more than my cousin Anne or Uncle Mutt. But in contrast to years living with Grandma Carney and Aunt Helen, I felt no emotional mother/child bond existed beyond the memory of her maternal helplessness shining in her sad eyes during the brief Logansport encounters years ago and short visits that followed a decade later. Earl Boyer was always kind and patient during my mother's worst episodes and saw that I was cared for by his folks during her many hospitalizations before I returned to Grandma C.

In all the places I'd lived: New Lexington, Logansport, Grandview, Upper Arlington, or Long Beach, not one of my peers ever raised the issue about my missing parents. The words "Crazy, Orphan, Stupid, Dumb, Retard" were not uttered by friends or my relatives. As I looked upon the broken face of my mother, I didn't harbor any ill will toward her. Nor did I look down upon her. She deserved my kindness like my cousin Anne or Uncle Mutt.

Misdemeanors & Felonies

Who I became as an adult wasn't solely due to the way my family raised me. In 1957, adolescents were indoctrinated into the world of heroes and villains depicted in newsreels, radio, movies, and TV. Any rites of passage for Upper Arlington boys were generally prescribed by white Anglo-Saxon Protestants or catholic parents of privilege. Children of the elite participated in private church summer camps, scouts, and Scioto Country Club's swimming, tennis, and golf programs.

Social graces were taught by Mrs. Pott's eighth-grade dance ritual called "Cotillion." Our families expected us to continue socially appropriate behavior long after we exited her program.

Where we attended high school set us on a specific life path. The public high school educational track was Upper Arlington Senior High. The private schools included Columbus Academy and the boarding Culver Military Academy. Catholic boys attended both educational tracks themselves. St. Charles and Aquinas across town were religious schools open to both boys and girls. Catholic girls often attended St. Mary of the Springs High School for Girls. All these schools provided education for the privileged, a group that I was part of due to my Aunt's taking me in, and I was about to graduate.

My last days as a Golden Bear were a wild ride best described as a youthful beneficiary of affluent white privilege in the period of utmost *freedom with a license* for delinquency, small crimes, felonies, and mayhem without consequences. Whereby my peers forged parents' sick notes to skip school and consume 3.2 beer and pizza. Spontaneous gatherings were held at my best buddy and classmate, Vincent Cardi's spacious third-floor bedroom which had two bedrooms and a bath. We'd hook up a jerry-rig stereo and blast Sinatra's, *Come Fly with Me*, Bill Haley's *Rock Around the Clock, Shake, Rattle and Roll*, and Everly Brothers' *Wake Up Little Suzie* and *Bye Bye Love.* School officials attributed the contagion of truancy to senioritis. Vince transferred to UASH from Aquinas High the same year I arrived. We called him, "Vinnie the Suave '' because he was the most handsome, impeccably dressed upperclassmen: parted dark hair, dark intelligent eyes, and member of senior scholarship and debate teams assigned to the "college track" curriculum. He was an officer in Hi-Y, active in dramatic clubs, and the Varsity golf team Captain who shared the Golf Championship with Jack Nicklaus the preceding year.

Vince was the second born of four brothers. Their father was a self-made gregarious and suave Italian immigrant who started a successful chain of upscale beauty salons located in high-end department stores like the *Union*. Everyone called Vince's mom, Miriam, "Mime*."* She was a natural beauty with lightly bronzed skin, and gracious. She took me in as one of her own even though she recently had resumed her goal of completing a graduate degree in psychology at The Ohio State University. Her academic ambition was a source of stress in her relationship with Mr. Cardi who identified with traditional Italian patriarchy. He spent time in New York City learning women's preferences in hairstyles. He loved jazz and personally knew band leaders like Bob Scobey and Louis

Armstrong that entertained at Columbus venues. Mr. Cardi was the family barber, cutting his sons' and my hair in the kitchen. The Cardis' had an *open-door* kitchen policy on weekends and summer. Mime would ring a cowbell announcing lunch or dinner. The first hungry animal racing in the door was a slobbering, female gray Weimaraner dog, Misty with vivid blue eyes, wagging her cropped tail, followed by a fat shaggy Tomcat. Then, the sound of feet hitting linoleum and a loud maternal voice conducting a head count: ***1...2..3..4..,*** her boys and stragglers flying in the room. Then she'd lock the door.

Unlike schoolmates George Collins, Bill Zig's, Art Wolf's, and Dale Smith's shiny cars and spacious decorated homes with maid and lawn services, Vince lived in modest circumstances. The Cardi's residence was an old three-story structure located at the corner of Arlington Ave, a stone's throw from the Upper Arlington City's Police and Fire Departments, and caddy-corner to an old strip mall which included *Kemper's Drug Store*, a two-chair barbershop, *Ohio State Bank,* a children's dancing studio, convenience store, and second-story apartments. The Cardis had a mammoth kitchen, and dining area with a long rectangular oak table that accommodated William and Miriam Cardi, their four sons, and as many guests. The living room was strewn with books. In my senior year, I crashed many nights here and at the Lamb's more than home. Vince managed to acquire a junkyard Chevy sedan with a rusted backseat floorboard, cracked windshield, and broken heater. Whenever kids bombarded it with snowballs, Vince stopped the car and I'd sit in the rear seat, retrieve a handful of snow through the floorboard hole, roll down the window and retaliate, laughing our asses off, reacting to their shocked faces. But unlike his car, Vince's wardrobe was first class: Oxford cloth button-downs, madras and herringbone sports jackets, fancy ties, Ivy league sweaters, *cordovan*, wingbacks, dirty bucks, and alligator emblem shirts; all were bargains from stores where his dad operated salons.

One evening Vince's dad took us to dinner at a fancy hotel downtown. Unbeknownst to me, after paying for the meal, Mr. Cardi said he had a surprise for me. He led us to the elevator that ended at the penthouse and motioned me to knock on the door. It was answered by a large Black woman beaming a smile. She wore a bright red dress.

Mr. Cardi said, "Jerry, say hello to Mrs. Louis Armstrong."

I don't recall what I said, only feeling strong arms hugging and then pulling me into a parlor teeming with red roses. Suddenly, I stood in a star-struck trance facing the Great Satchmo dressed in a tuxedo, loose suspenders, starched shirt dangling bow tie. My trance was broken when he reached out, hard-pressed my lame hand until it hurt; his smiling lips revealing the artist's trumpet-shaped callous mouthpiece blowing genius, snow-white teeth in the cavernous mouth, marble-size eyes glinting Black's dark history swallowing me whole; his raspy voice rattled my brain senseless.

There was an inverse relationship between the number of days to graduation and incidents of mayhem at school. Ben Tobin, a burly senior football teammate that enlisted in the Marine Corps Reserves, and I flooded Mr. Cavin's chemistry room by connecting *Bunsen* burners to water hoses. A week later, Dale Smith, Trent, and I replaced Old Glory flapping on the school's flagpole with bras and panties. When the flag was raised again, we were ready. We cruised country roads in Dale's mother's new yellow *Mercury* convertible searching for roadkill. The next school morning, hundreds of students' eyes locked on the carcasses of two cats and a squirrel. The same week, the senile history teacher graded my term paper, *Evolution of WW I Armored Tank*, plagiarized from the encyclopedia in Dr. Watson's library-"B- too bookish." He scrawled on the top.

Ha, I thought to myself.

In the ultimate irony, on "Government Day" I was selected as the acting Upper Arlington City Police Chief For A Day participating in a ride-along, twirling a baton and cuffs, inspecting riot gear and guns, observing a lineup of hoodlums at Columbus P. D., observing a male juvenile probation officer counseling a 15-year-old delinquent boy caught stealing a car. I only had two negative personal experiences with cops. When I was seven, one of the U.A.'s finest roused me out of a vacant lot's tree fort in progress and took me home by the collar. Aunt Helen chastised the cop for "manhandling a child." In the second incident, I was 17, walking home at midnight from a girlfriend's when a pair of Upper Arlington cruisers stopped me on the street. Two officers jumped out and threw me across the hood. One asked what I was doing there. My face pressed against a hot metal hood over an idling engine, I said, "Walking home."

"Your address? Gotta ID, kid?"

"Driver's license," I said, turning my head.

He released one arm. I retrieved my wallet from my hip pocket and handed it to him. He gave it to his partner holding a lighted flashlight.

Seeing the address, he motioned his partner to release me. The officer stepped back, eyed the prominent gold **A** emblem sewn on my black letterman's jacket, and returned the wallet. I asked why they stopped me. He pointed at the residence three houses from the Watson's. A police cruiser was parked in the driveway.

He said, "The owners reported a break-in. The only item missing was the family's iron safe. The robbers broke a rear window." The cops didn't apologize. I walked home, reflecting on the Watson's large steel safe hidden under a tablecloth, next to my cousin Jane's four-poster bed. Figured the thieves must've had prior knowledge of the neighbor's safe and family schedule.

Carlie Will was the most popular and inspiring high school teacher. He taught English. He'd sit cross-legged on his desk and read quotes from James Thurber's *Winnie the Pooh:* "Always watch where you are going. Otherwise, you may step on a piece of the forest that was left out by mistake."

I related to him that the Thurber quote reminded me of my Uncle Mutt's country lessons: "Walking pastures: Never step twice in the same pile of shit." Charlie let the class present my cowboy adaptation of *Hamlet*. He introduced us to Emily Dickenson and Walt Whitman. After reading *Leaves of Grass,* I told him, according to official historical U.S. Army records, one of my maternal Carney clan ancestors-Michael Carney- was enrolled in the Union Army Co. G-61- Regiment of Ohio on the 2nd day of March 1862 at New Lexington, Ohio, He volunteered to serve three years, survived the Battles of *Gettysburg (*July 1-3, 1863), and *Lookout Mountain* (November 24, 1863). In the latter, he served the 11th Army Corps in V.A, Old Joe Hooker's 20th Army Corps. It was said, he was so grateful for surviving the war that he seriously considered entering the order of the Preachers as a Dominican Brother; the same Catholic order my cousin Jim Watson was eventually ordained to, a never-ending reward for his mother's prayers and dreams. I confessed to Charlie Will that with the single exception of my second-grade teacher at Lowell Elementary in Grandview until she was fired, no adult ever read a children's book to me in my childhood. Students associated Charlie Will's exaggerated gestures, soft voice, and feminine mannerisms with homosexual orientation. But he was never disparaged because his readings opened long-repressed voices of adolescents' imagination and quest for truth and acceptance.

The most serious illegal incident involved breaking into school to steal final exams. Traditionally, the administration waived the final exam of graduating seniors in courses with a B average. The incentive motivated students to study extra hard and qualify for the waiver, thus skipping exam week. All was working out as planned until the civics teacher caught me wandering in the hall without a fucking "hall pass." He was like a truant officer lurking behind a tree or in the shadows waiting to catch the school's notorious, yet never held accountable prankster. Not about to do me any favors, with a shit-eating grin, he meted maximum punishment: I would have to take final exams. I gave him a cold stare, plotting my revenge.

I dreaded the thought of nights cramming for tests and the lost pleasure of peeling paper labels off 3.2 Stroh's beer bottles consumed with sidekicks at the neighborhood *Arlington Grill.* Providence intervened. Ben Tobin, our team's toughest lineman, and the only classmate that joined the *U.S. Marine Corps Reserves* had something in common with a mutual close friend, Trent Lamb. Both were borderline students whose diplomas hinged on passing all finals. Ben dated the school janitor's daughter. She loaned him the master key to all the rooms including the mimeograph room adjacent to the principal's office. Friday night before exam week, Ben, Trent, and I sneaked into the test vault and gathered copies of all relevant exams; pledging to distribute tests only to a few trustworthy friends with the caveat: no grades above 90 to avoid raising suspicion.

During the post-exam week, while I was sitting in Homeroom celebrating in my mind the success of our exam caper, the classroom speaker suddenly cracked a voice summoning all seniors post-haste to the auditorium. Students slowly filed in thinking Principal Tojo was about to discuss the graduation ceremony or present awards to brown-nosers. This thought went up in smoke when I saw one of Upper Arlington City's finest standing in view off stage. And Tojo, who looked like a stand-in for America's WWII Japanese nemesis in a suit, walked across the stage to the podium, paused for silence, retrieved a hand-held microphone, scanned his charges through thick lenses of his round wire-rim glasses, then announce: "All seniors' final exam results were declared invalid because analysis of test results revealed an abnormal pattern of high grades that could only be explained by test takers having prior knowledge of answers. Consequently, I have instructed the faculty to revise exams and reschedule the tests for next week."

"Fuck me," I muttered, sliding down in the seat, eyeing my fellow culprits, anxious over the threat of cops taking fingerprints, which probably would have been routine if there was evidence, that we'd "broken" into school or jammed the mimeographed room door lock. The master key was ingenious. Next time we'd wear gloves.

Enter plan B; providence intervened again. A student scholar- exempt from finals-who assisted one of his teachers in revising a test told his girlfriend who passed it on to her girlfriend who confided in me, heard that Tojo ordered teachers to hide new finals in their respective classrooms, not in the mimeographed room. No problem with a master key. Wearing rubber gloves, we copied freshly mimeographed tests, inhaling ink. Unlike before, we honored our pledge not to share answers. This time, the teachers were fooled, and I graduated along with my close friends.

My high school graduation gifts from the Watsons were a suitcase and *Timex* watch counting hours and days of changing landscapes. Neither of my cousins attended. Jim was ordained "Father Vincent" guiding his flock in a small town in Ohio, and Jane was away visiting college friends. Meanwhile, their parents, or the soon-to-be empty nesters, were building a ranch house in Upper Arlington's expanding community. The Onandaga house was placed on the silent real estate market brokers used to sell upscale homes restricted to white gentile buyers. Pending construction completion, we moved into a three-bedroom upstairs condo on Arlington Avenue, a block from the Cardi's residence.

The importance of attaining top grades, preparing for college boards, or meeting admission standards of tier 1 institutions of higher learning was never discussed at home. I brushed off the prospect of a football scholarship without Coach Morehead's recommendation. Returning to the previous summer's concrete job was not an option. The country was in the depth of the *Eisenhower 1957-58 recession*. My ex-girlfriend's father's construction business went bankrupt. Thousands of Ohio workers were being laid off. Trent worked part-time at his mom's advertising agency. Cardi got a clerical job with the state through his dad's connections. I earned pocket money fertilizing seedlings and trees at Watson's new property. This involved driving a metal stake into the ground with a sledgehammer to make holes for fertilizer. I also organized a group of female ex-classmates to solicit neighbors to pay us to paint home addresses on curbs for fifty cents. Few took us up on this once-of-a-lifetime opportunity.

Summer Caper

Water swirling, spinning us around

from honesty

to lies,

and back again

I was lucky to survive.

On the first July weekend, Vince and I drove his youngest brother, Phillip to summer youth camp near Lake Erie. After dropping him off, we decided to cool off, wading barefoot in Lake Erie. Suddenly we felt our bodies sinking into the dark abyss of a drop-off, instinctively treading water, and madly paddling to keep our heads above water, fighting the downward drag of waterlogged jeans and T-shirts. A pitiable swimmer, watching Vince bobbing beside me and struggling to remove his shirt, I contemplated death by drowning until I felt my raw toes clawing the sandy bottom, my body moving in slow motion back to earth.

We lay supine exhausted in wet jeans on the beach in silence, gazing at the twilight sky. Our dry sneakers and socks lay a few feet from us where we left them. I assumed we'd get a motel room for the night to recuperate and return to Columbus the next morning. But Vince had other plans. He reacted like a victim of a violent crime, demanding restitution, and revenge. He said, "Fucking lake stole my graduation *Bulova* watch and *Alligator* shirt. It owes me."

I joked, "Thanks to the leather band, my *Timex* took a licking but keeps on ticking."

He related a buddy in his old neighborhood's fighting philosophy. "Win or lose, never leave a brawl without something of value: a ring, jewelry, watch, wallet, piece of the ear, or broken tooth."

I retorted, "Vince, what are you going to beat out of Lake Erie, dead fish, a piece of driftwood?"

Searching for a motel, we passed a Girls Scout camp along the lake. We saw girls dressed in shorts and swimsuits, busy building a campfire and erecting tents. Slowing down, Vince's eyes locked on a dozen aluminum canoes and flatboats moored by rope to a nearby pier.

He said, "Jerry, look for a hardware store."

"Why?"

"A car rack for a boat."

"Are you shitting me?"

The clerk led us to a display of life preservers, boat seat cushions, ice coolers, gas cans, canoe paddles, tie-downs, water-proof bags, heavy-duty flashlights, and flare guns.

"That'll work," Vince said, pointing at the steel bar car rack and rope.

"Jesus," I said, assisting him in securing a rack on the roof of his family station wagon.

Following dinner at a greasy spoon and four hours of restless sleep at *Motel 6*, we drove to the outskirts of the girl's camp and hiked in the moonlight toward the pier 100 yards away. The campfire was reduced to ashes and embers reeking burnt pine. All tent door flaps appeared secure. The sounds of rustling leaves and moored canoes and *Jon boats* clanging in the wake of the lake's current masked our movement.

Wearing sneakers and shorts, we slipped into the cool water, wadding our shoulders to the tethered fleet. I untied the rope of a two-man flatboat moored to a pile, climbed aboard as Vince retrieved two paddles, and joined me. Stealth as commandos we paddled to shore. We dragged the prize to the Chevy station wagon, silently lifted and lashed down a hundred-pound 12-foot flatboat on the car's roof. Neither of us had any intention of using or selling it. To avoid detection, we drove straight home at night, unloaded the boat in Vince's driveway then carried it to the backyard. The racks were stored in the garage.

Sitting on Vince's back porch, sipping 3.2 *Rolling Rock*, gazing at zinnias blooming in the bow of the aluminum Jon boat that Miriam Cardi boasted was the "flowerpot in Upper Arlington," it hit me. Suddenly my body cringed. In the eyes of the law, we were adults that could be charged with a felony crime. Add the final exam caper of breaking in, previous juvenile acts of vandalism and petty theft, and the image of joining convicts marching in lockstep to the prison chapel and living like an animal in a cage, flashed before me. More significantly, the feeling of deep shame over violating the memory of Grandma Carney's unconditional love, selfless sacrifices, and unwavering loyalty for the Carney's wounded and strays, and the risk of disgracing Grandma's oldest daughter Helen whose timely interventions saved a runaway orphan from the fate of semi-illiteracy, possible reform school, and regimented army life. I silently vowed to leave my life of crime.

August's conversations with my friends turned to college. Vince and Trent appeared to be heading to Ohio State, a short ride from home. My college future was not mentioned at the Watson's. I explored ROTC and U.S. Marine Corps scholarships that commissioned graduate's 2nd Lt. obligated to serve four years of active duty. I visualized myself in an officer's uniform, being a fourth-generation member of the Carney clan of Perry County Ohio whose ancestor Michael Carney, was a Union Army volunteer that survived the Battles of Gettysburg.

The first hurdle to qualifying for the Marines Platoon Leaders program was passing a comprehensive physical conveniently arranged by my Uncle J.O. at his hospital. I left the hospital confident of passing both physical and future college courses because my study habits had improved. "Pass to Play '' unwritten rule-governed sports eligibility in parochial schools and UASHS. I rejected any thoughts of failing college work and the physical challenges of the Marine Corps.

Fuck me. I failed the physical. The urine test showed high levels of albumin associated with alcohol consumption. The night before the exam, Trent and I got shitfaced guzzling Stroh's.

Vince's folks were so opposed to me volunteering for the draft that they offered me a place to live and attend Ohio State with Vince. Mrs. Cardi advised that I asked the Watsons, whom they never met, for financial assistance to attend college at Ohio State. Mr. Cardi said I could earn money cleaning his salons. I had never breached the subject of needing college assistance with the Watsons. College applications of friends were in the mail or being processed. Amidst packing precious China and antiques, I sat at the breakfast table across from Aunt Helen, sipping coffee and smoking a *Pal Mall*. I was anxious about raising the issue of college support. She paid me for chores and an allowance during my school years, and I was always resourceful earning pocket money in the summers. I told her I was mulling over applying to Ohio State and accepting the Cardi family's offer of free room and board and working part-time for Vince's dad. Eyes squinting from chemical fumes exhaling nostrils and cracked mouth, she took a long drag from the cigarette, turned her head toward the window overlooking the block the Cardi's residence, then leaned over a green translucent pipe astray, smothered her addiction in the graveyard of half-smoked unfiltered cigarettes marked with red lipstick. She removed a piece of tobacco lodged in her teeth.

"Honey, Ohio State is out of the question. We're thinking more like John Carroll University where Maggie Sullivan's son Mike is going."

Mike was a St. Agatha straight-arrow classmate. Reading my compressed lips and determined eyes she slowly relaxed her posture, eyes glinting with a hint of softness not seen since our Greyhound Bus Station reunion.

"Well, we helped you get this far and will continue, but on the condition that you go to a catholic college. Frankly, J O. expressed doubts about you cutting it. He has never seen you studying at home."

I detected from her tone that Ohio State failed two important criteria: it was geographically too close to home and a public institution of higher learning. Predestined for the priesthood, Jim attended Notre Dame, and Jane followed her mother's prescribed path, going to St. Louis University.

The thought of being taught by Jesuit priests was worse than joining the army. Aunt Helen was strong-willed and uncompromising when it came to the Church, and few could match her pride reflecting her social station. The latter was her Hercules heel.

"Well, I guess my only college option is accepting help from the Cardis."

Suddenly she set up in the chair, retrieved, lit, and inhaled a cigarette, smoke rising fading cloud over the kitchen table, eyes blinking eroding social station and control. She knew she met her match in perseverance and independence. Her children always caved. After a long silent standoff, Aunt Helen amended the offer to pay for books and tuition to an institution of my choice that accepted me, "excluding Ohio State" of course.

Dumbbell English & Milk Runs

Moving back to the beach,

this time, different

I was an adult

Working, schooling, and playing football.

I was Ernie's equal

not his ward.

Safe.

Long Beach State was a few blocks from the Paskell's. When I arrived, I remembered again that, according to Grandma Carney, Aunt Marg was my mother's closest sister, and always nice to me despite Ernie's volatile personality and his regular child abuse. Perhaps Ernie adopted new ground rules since Aunt Helen failed him as her orphan nephew's guardian. In the future, I would be an adult guest, not a subservient ward.

Founded in 1949, Long Beach State Liberal Arts College encompassed a sprawling landscape along N. Bellflower Blvd and E 7th Street. The *upper campus* included an administration building, library, theater arts auditorium, classroom buildings, and Student Union. The *lower campus* housed a co-ed dormitory, state-of-the-art basketball gymnasium, locker rooms and water polo swimming pool, football practice field, tennis courts, archery range, vocational education classrooms, vehicle maintenance garage, and campus post office. According to the college catalog, per semester tuition for a full-time student was less than $100 for a *resident* and $400 for a *non-resident*. Since I graduated from an Ohio elementary and high school, I assumed the Watsons would cover the cost of a *non-resident* and textbooks. The open admission policy of California state colleges entailed two days of grueling academic tests to determine applicants' proficiency in math and English. BA students requiring *remedial* assistance had to complete *Dumbbell English.* It was universally understood that open enrollment public colleges used *English Composition* as a screening tool to weed out Blacks and Hispanics.

An LBSC advisor flipped through my test scores and recommended declaring social science as a major. He said, "Gerald, don't be insulted taking *Dumbbell English*; half the freshmen here are in the same boat. My nephew had to repeat the damn course to get off probation. But, if possible, avoid Dr. Libby. She's a stickler for grammar: three strikes you're out."

I tried out and made the first-string freshman football team, coupled with two years at Long Beach St. Anthony Catholic High school got me registered as a resident student. I joined *Sigma Pi* fraternity because it was composed of commuter students lured to the off-campus fraternity house across the street from the beach. On weekends, the house mother was off duty and the dining room featured dirty movies and slideshows of naked women taken by GI and Navy vets who "toured the world." Every floor of *Sigma Pi* house had a Coke machine that dispensed *Olympia* and *Hams* beer for a quarter a can. My fellow pledges and future frat bros included Jack Wilkerson, Nick Gordon, Jim Bright, and original Walt Disney Mouseketeer and Lawrence Welk Dance performer, Bobby Burgess.

Facing academic probation in Professor Elizabeth Libby's *Dumbbell English* class, I sought out a *Sigma Pi English Lit* master's student Tom Martin who edited my take-home composition essays and tutored me in writing. He was a Korean War combat vet that survived the *Battle of Porkchop Hill.* He volunteered that his worst experience was seeing a GI shot by a fellow soldier. Tom said he was among the last group of walking survivors following behind litter bearers carrying our wounded down a steep hill of loose rocks, bomb craters, Chinese bodies, and jettisoned equipment. When the first litter bearers reached the bottom of the hill, the GI they were carrying jumped off the litter and said, "Thanks for the ride." Suddenly, one of the exhausted litter bearers, dropped his end of the litter, unslung his rifle, shot the malingerer dead, and left the body where he lay for graves registration to retrieve. No one blinked an eye or said a word.

I shared Ernie's story of- "submarine justice," telling Tom about a sub crew throwing a despised officer overboard to certain death-during war at sea.

Jack Wilkerson was a starting freshmen team's offensive end. Jack lived at home with his stay-at-home mom, younger sister, and dad who worked as a welder at the Long Beach Naval shipyard. Jack drove an old English compact car of questionable age, missing bumpers, donut tires, and a rusted body encrusted with barnacles from years parked along a road exposed to the sea's air. He paid a Pike Amusement Park carny $190 for the car with money he saved flipping burgers at a greasy fast food eatery. Each shift of gear raised the pitch of tortuous whine assaulting the ears. Like me, Jack was suffering psychic and physical withdrawal from the scent of sweat, grass, dirt, and pigskin, and the aggressive release only field combat provides. To assure his draft deferment wasn't revoked we registered the minimum 12 credit hours. I ran laps and memorized the playbook. By the third practice, I was starting fullback on the freshman team. Recalling my aversion to late hits, I stomped a tackler in scrimmage; aluminum-tip cleats impaling his leg. I had to wash the blood off my shoes. Poor bastard spent six weeks hobbling to classes in a cast. I opened a $350 personal checking account thanks to qualifying as a resident, a venial sin I committed when I failed to report my tuition savings home, rationalizing I saved the Watson's thousands in rejecting room, board, and tuition at a private Catholic college. Used hardbacks: *Introduction to Economics*, *Cultural Anthropology,* and *Business Law* were still outlandish. Paperback novels for *Introduction to English Literature: All Quiet on the Western Front, Of Human Bondage, The Ambassadors, and Tom Jones* were reasonable on the pocketbook and merited a place in one's library

The freshman football schedule pitted us against junior college players that had varsity potential at four-year schools. The most attended 1958 LBSC freshmen football game was held mid-season at Mount Palomar Community College at San Marcos; elevation of 1900 feet. The *Mountain-peak Stadium* offered every amenity to visiting teams and hungry fans except fucking oxygen. Our numb-nuts coach enforced calisthenics and a warmup that left us gasping for thin air before the first kickoff. It was the last time I heard my name announced, "ball carried by Wheeler" over the public address system. Bolting through the defensive line into the secondary, I attempted to straight-arm a tackler. Suddenly, I was blindsided out of nowhere, the combined weight of my adversary's and my body falling face down, jabbing my extended arm into the hard ground. What followed in lightning flash speed was the sound of referees' whistles, the cracking voice of the public address system, and muffling my painful cries resulting from an arm savagely ripped from the shoulder socket.

No x-rays. No-pads practice for a week.

The team's trainer immobilized the right arm, taping it to the side to prevent it from aggravating the injured shoulder. Like a one-wing gooney bird flapping its wing, my flying days were over. A week later, I was lying in the grass outside the Student Union, gazing down at fellow archery class students walking to their target stands to retrieve arrows. Hell, I couldn't even string a fucking bow. The football coach was so pissed after I quit, that my photo and seven touchdowns were deleted from the *49er 1958-59 Yearbook.*

I didn't regret my decision to quit. Long ago, I concluded that recreational and organized sports are governed by a merit system with no excuses or favoritism. Because I left the team after an injury, my excuse for leaving pushed me out of the Coach's favor.

Yet, I had learned from my many years of football and my numerous guardians about how to achieve success. Success, they taught me, was attributable to talent, endurance, perseverance, practice, practice, and a positive mental attitude. All skills homed in my passed-around existence while learning the meaning of family loyalty, empathy, and interdependency of friends and caretakers would help me figure out life. My role models showed me the value of hard work. Grandma C cooked, washed, and patched clothes, and nursed my wounds; Grampa C shucked corn, peeled homegrown spuds, and carved toys on the kitchen steps. Mutt was a principal mentor in taking care of the small farm and livestock and mining coal that supported our livelihood during hard times. Indeed, I was the beneficiary of protective and non-judgmental Carney clan caretakers that instilled a sense of self-pride and growing self-confidence in carrying out chores and testing boundaries.

Mental illness and physical disabilities in the family taught me acceptance. In Logansport, I was both dependent and caretaker confronted by the ravages of my mother's broken mind and mood swings ranging from self-destructive manic behavior to catatonic withdrawal. I was unfazed by circumstances that catapulted me from second-grade classroom to troop train bound to Columbus, back to Grandma Carney and Uncle Mutt where I bonded with 9-year-old cousin Anne Carney born with Down syndrome who lived next door. She shared her toys, TV, and dreams and defended me from neighborhood bullies. I was her shadow like Uncle Mutt hoovered over me at the Carney homestead and abandoned strip coal mine. Over the years I wondered what happened to Anne, her mom Evelyn, now a widow, and Uncle Mutt. I missed Anne and my Uncle Mutt, even as an adult. That they suffered from physical and mental disabilities didn't matter one bit.

I didn't miss my mother. It had been months since I saw my biological mother Kathleen rocking compulsively on the park bench at the State Mental Hospital, smoking a hand-rolled cigarette, waiting for her son's visit. What happened to the California duo brothers Aunt Helen conscripted to transport me to my exile and the Navajo woman in beaded buckskin dress at a Route 66 trading post who sold me a *Dream Catcher* that fails to catch haunting nightmares? These people barely entered my life, but they impacted me in so many ways.

Ernie quit the *L. A. Times* route when Bill accepted a summer job at *Knotts Berry Farm*. Bill had grown into a lanky teenager with zits heavily populating any real estate on his face. He parted his blond hair almost in the middle like his dad and chain-smoked *Marlboro's.* Bill emulated famed film director John Ford's image of a wrangler or gunslinger: dirty jeans, western shirts, and boots. Poster images of Johnny Cash abounded on his bedroom walls. When Bill was home, the sounds of the deep bass, baritone country voice, and strumming an acoustic guitar of his singing rebel idol reverberated down the hall: *I Walk the Line, Big River, Don't You're Your Guns to Town, One More Ride, Ballad of Teenage Queen.*

Ernie replaced the 52 *Ford* with a 56 *Chevy Bel Air* station wagon. Instead of throwing the *Times* before dawn for extra cash, he loaded a friend's milk truck with *Elsie Borden's Multi-vitamin* and cottage cheese at a local dairy. By 8 a.m., he changed from leisure denim jumpsuit into dress slacks, white shirt, and tie then drove a station wagon packed with *Bostitch* stapler supplies including compressor-operated heavy-duty staple guns used to upholster furniture. Meanwhile, Aunt Marg had been promoted from bank teller to a loan officer and drove daughter Marjean, now a shy preteen blonde, to St. Bartholomew Elementary in the *Chevy* convertible. Bill drove the 52 *Buick* to high school and *Knotts Berry Farm* where he worked weekends. The unspoken punitive side effect of my running away was being denied access to Paskell's family cars. They'd drop me off at campus if it was convenient. But, like the days living with the Watsons, my transportation was dependent on friends, bus, feet, bike, and thumb, in that order.

One Sunday I accompanied Bill to *Knotts Berry Farm.* I asked him, "How did ya' get your job?"

He said he heard they needed help for the stagecoach show, one of the popular summer venues during the peak summer season. He admitted he did not know shit about horses. According to Bill, the venue schedule called for a stagecoach on the trail, and another on the stand. Each stagecoach had a supporting crew of a driver, a shotgun rider, and a cast of dozen white actors playing Indians, cavalrymen, highwaymen, and passengers. *Knotts Berry Farm* maintained a stable of horses overseen by a barn manager, groomers, and handlers. New hires must demonstrate riding skills or complete riding lessons. All start at the bottom of the pecking order. First day: the barn boss hands you a shovel and points to a pile of manure. Second day: clean horse tack bath and groom horses. On the third day: paint your faces like a Sioux warrior, dress in buckskin and moccasins, ride a horse bareback, whoop, and holler chasing down a stagecoach, then grab a purse from a screaming poor pioneer woman passenger until she's saved by the cavalry. Next week: wear chaps, boots, and spurs, and cover your face with a red bandanna. Hop on the saddled gelding, then gallop and empty six gunshots into the stagecoach driver and shotgun rider and steal the strongbox filled with paper money. Then it will be your turn to play a *Bluecoat* blowing trumpet charge. The highest ranked and most desired summer position, the one that required the most skill, was the stagecoach driver. The second in desirability was playing a shotgun rider carrying a double-barrel shotgun loaded with blanks. These two positions were interchangeable but exempted from other roles.

Bill was proud to have moved up to stagecoach driver by the end of the summer season and kept hold of the reins on weekends after school started again. I related to Bill that my most painful experience riding a saddled horse was being thrown and dragged on gravel pavement by a spooked rented gelding which scraped my ass so raw I had to wear Kotex and a bathing suit to conceal blood seeping through my jeans until it healed. Bill started each show, dressed as a young cowboy; he tipped his wide-brim riding hat at a pretty pioneer lady, extended his hand to grasp a ticket, and said with practiced charm: "Mornin,' ma'am. I'm the stagecoach driver. Welcome to Knott's Berry Farm." I volunteered to be a stagecoach passenger robbed by an armed, masked highwayman.

Upon returning to the barn after the show, I assisted a groomer in unharnessing a lathered team. I observed my cousin brandishing a bullwhip, standing outside a stall of a recalcitrant colt. Suddenly, I saw his eyes flare with anger just like his father's eyes flared the day he viciously smacked Marjean for ignoring his order to bring him chocolate syrup for his ice cream. Like his father, Bill's lips clamped in sync with gunshot cracks of bullwhip making its mark on the haunches of the yearling until it complied with his demands.

I mailed the first semester grades to Aunt Helen, not mentioning the football injury and tuition savings of qualifying for resident status, the anxiety over nearly failing dumbbell English composition with a *D,* and suffering through *Principles of Economics* and *Speech-Voice-Diction* with *C's.* The latter was taught by a tall handsome ex-Marine who lost a lung at *Iwo Jima.* The feeling of satisfaction and triumph in overcoming initial trauma over *Blue Book* exams and feeling grateful to be saved by a dictionary had me not minding the lower marks. Besides, in *English Literature* and *General Physical Anthropology*, I earned *B's*. A Christmas holiday job selling men's clothes at *May Company* sustained me financially until Dr. Watson's spring semester largesse. My social needs were met through fraternity and class relationships. Weekend parties at our Sigma Pi house and UCLA and USC chapters were common. But Monday through Friday, it was all business. No booze. I lived in the library, studying alone in a carrol or soundproof booth listening to 33 rpm records like Gershwin, Raskonokolf, and

Richard Rogers *Victory at Sea.* I'd outline and re-outline class notes and textbook chapters writing in smaller and smaller letters until the contents filled a few legal-size pages. Peers often treated me as if I was smarter than they were rather than attributed my higher grades to late-night hours of study. I read literature on love, hate, and war's slaughter: handwritten obituary, *I died June 12, 1864, Cold Harbor* pinned on dead Yankee uniforms, *Andersonville*, *Trail of Tears*; doughboys trapped in no man's land at *Somme* under artillery bombardment and machinegun fire while brothers in arms hunkered down in a maze of trenches flooded with blood, piss and shit and rats feeding on human body parts in open graves; Nazi and Japanese atrocities: *Holocaust, Malmedy Massacre, Pearl Harbor, Bataan. Lest we forget Dresden, Firebombing Japan, Hiroshima, Nagasaki, and The Execution of Private Slovik.*

I had no idea where a college education would take me. I recognized the material and social advantages of living as a privileged white kid in a prominent doctor's family where the first order of business was education. Notwithstanding adventurous early childhood years doing farm chores and exploring coal mine shafts that only mined grief, living in a house surrounded by books and magazines on nearly every subject and country of the world was a prescription against boredom on rainy days. The Watsons were the last family to purchase a TV. However, unlike most of my Upper Arlington peers, I had the *Roosevelt Recreational Park* perspective, interacting with Columbus Eastside working-class, rural, and welfare kids surviving on the skills of the street: scavenging for toys, relying on makeshift bikes, running errands, and collecting deposit bottles for pocket money. No weekly allowance or personal bike at Grandma Carneys. Lugging golf irons at age 13, then California exile, hard labor home improvement servitude and dragging water-filled paddleboards in the wet sand at 14, runaway at 16, pouring concrete in the summer heat at 17, studying the faces of men with dead-end jobs. But I knew these jobs were only temporary for me. I relished the shared self-directed adventures of unspoiled kids and a few sons of privilege. Fortunately, as a Watson's ward in the early school years, non-academic activities were self-prescribed: sports, river, and woods. I felt a deep sense of mission in not disappointing my benefactors, especially in honoring the memory of my Grandma Carney's sacrifices, grit, and overcoming a life of grief.

The Watson's spring semester tuition check, mailed from his business office, carried me through March. By the time I ran out of money, all student campus-based student jobs were filled. Fortunately, Ernie's *Bostitch* stapler sales had increased, allowing him to share the two-hour morning milkman job that paid $15 a load. I'd rise at 4 a.m. Monday, Wednesday, and Friday, walk a short block to the milkman's house, pick up an empty milk truck, drive three miles to the dairy then stack dozens of heavy cartons of quart-size glass bottles filled with white *Elsie Borden's Multivitamin* or chocolate milk, pints of half and half, and paper cartons of cottage cheese behind a wire barrier behind the driver. I drove half standing up, double-clutching grinding gears and listening to yapping dogs, rattling bottles, scanning the road for potholes and road hazards ranging from trash to construction debris. I'd sip chocolate milk, visualizing a tall blonde pretty student actress named Linda who I was working up the courage to ask out. To stay awake, I'd slap my face, and recite lines of *Long Days Journey Into Night, and A Streetcar Named Desire* for drama class. All went smoothly until the morning I heard the sound of an exploding tire, felt the truck list out of control, and crashed into a fire hydrant. My body slammed against the iron dashboard. With wide eyes, I gazed at the shattered glass and felt the white sea geysering my future.

Oil Pumps & Solvent

Body rocking, pitching, lungs burning, breathing dust, diesel fumes, I found myself ascending winding dirt roads reeking of oil, my hands fighting the steering wheel of a pickup to earn fall semester's pocket money. I scanned the mountainous landscape under siege by an army of pump jacks extracting America's addiction into storage tanks out-of-synch with planets and stars. Braking then coming to a stop by a drill pipe suspended on a giant mobile crane dripping black ooze from the hazy California summer sky, I was surrounded by a gang of leather-face, barrel-chested roughnecks, tool pushers, and roustabouts with massive shoulders. These were big white men with black oil faces wearing grimy dungarees, khaki work shirts, laced-up high-top boots gleaming with both mud and crude oil,] like a bronze memorial to their breed. They proudly doffed dented tin hats, revealing skull gouges, nicks, and scars, gesturing with mangled hands: the legacy of flaming derrick timbers, burning oil, falling tools, pipes, flying chains, and body parts of forgotten casualties of oil well explosions. Narrowed eyes took inventory of my puny physique and youthful demeanor; they flexed their bulging arms tattooed with surreal images of eagles and naked women.

I felt grossly out of place. To their credit, they never said anything outright. They paused, spit tobacco at fire ants evacuating volcanic nests, and gave me "the college boy treatment:" A big brute stepped forward, slapped a hard hat on my burr head, punched my shoulder almost knocking me over as fellow roughnecks retrieved and secured a disabled oil well pump rod and valve assembly in a vice welded to the truck bed. A husky voice said, "Kid, I hope ya know whadafuckyadoin,' cause the drill man hates paying idle crews."

Like James Dean in *Giant,* I took a deep breath, grasped a heavy shark-tooth wrench and pipe "cheater," and tightened my grip on the pump assembly. Standing on tiptoes, I raised and pulled the wrench down with both hands until my muscles screamed, and I heard the metal joints crack like a snapped bone, hoping it was not mine. Then as if rehearsing for a stage performance, I dipped parts in a coffee can of pungent solvent, removed the steel ball from the socket, and inspected it in sunlight for pockmarks like it was a mini-ball relic of the Civil War *Battle of the Wilderness*. Next, I practiced procedures: I reset it, gave the valve the "blow test," and listened for a flaw– spitting sand and oil. Wiping away oil and dirt on a silver potassium pump with a rag soaked in a solvent, I could measure the wear of a million strokes with a micrometer screw gauge. After a dramatic pause, a pause I learned in acting class, I give the hunched gang in tin hats the diagnosis they did not want to hear: they need a new pump.

Groaning, the men took a break. I sat with them. Blowing smoke in my eyes and ears, they complained about their disloyal wives, ungrateful children, low pay, cheap "three-dollar- a- barrel oil," warning about the "bust to come." I sat and nodded.

Driving back to the shop to retrieve and return with a new five-thousand-dollar pump and assembly used for high-producing wells, I thought about the worst jobs of my passed-around existence. I detested the freezing face and hands earned from throwing newspapers to the porches of ungrateful homeowners, occasionally performing the task while riding a bike in blinding snow. The exhausting job of pedaling an ice cream cart uphill in scorching heat was not any better. Even more tiring and punishing on a body was the job of hauling rocks, laying flagstone, and weeding dichondra grass on hands and knees then digging ditches for a sprinkler system. The cyclical nature of this job, creating well-manicured lawns, so more misery will grow was asinine. My future wasn't in caddying duffers' double bags– 18 grueling holes for lousy tips– nor was it working concrete in dank basements. Striking oil two years later felt as if I won the summer job lottery--*Oil Well Repairman* at *Pacific Pumps* earning $3.50 an hour. How did it happen?

Gaining that job required a bit of luck and some skilled bullshitting. Thinking I was not taking adequate initiative in my job search, Ernie drove me around Long Beach scanning strip malls for posted "help wanted" signs. Fate delivered us to a nondescript building at the bottom of *Signal Hill,* a small mountain landscape covered by oil derricks, jack pump wells, and storage tanks owned by Shell Oil Company; this was the very oil company found liable for the 1954 Long Beach oil tank explosion that rained oil on the Paskell's house and neighborhood. A man-made disaster, this event resulted in the torment of wiping oil stains off the wood with turpentine in the morning and painting during the afternoon throughout the first summer days of my exile.

A middle-aged man wearing a white shirt and tie and slacks answered the door. He was holding a stack of papers. I asked if he needed help. Before he responded, the phone rang. Behind him, the muffled sounds of moving machines or tools in the back of the building stretched out to the switchback gravel road leading to the summit of Signal Hill, hummed faintly. He motioned me to wait as he double-timed back to the office to answer the phone. The knotty pine front walls were decorated with historical photographs of wooden oil derricks gushing oil, graphic posters of oil well equipment and pump parts, and a Marlboro cowboy wall calendar. Eyeing Ernie puffing a cigarette in his car's open window, listening to Vin Scully calling Dodgers' hits and errors, I was about to give up when the *Pacific Pump* man waved me into his office. I anticipated he'd ask about my last job experience-"crashing milk-run."

I don't know what I would have said in response to that question.

He turned to me, paused, and started speaking. With his slumped shoulders and filthy hands holding paperwork, his blank green eyes looked me up and down. "Kid, you got no fucking business taking up space at my office or my valuable time or scintilla of skill to reduce my burden...., kid, can you type?" he demanded.

"Sixty words per minute on blind keys, taught by Holy Cross Brother at St. Anthony's High," I responded.

"When can you start?"

"Now."

He sighed, placed the paperwork on the desk, and extended his hand. "Doug Matthews. I manage this branch of *Pacific Oil Well Pumps."*

I followed him to the shop floor at the rear of the building. I observed two men, one dressed in oil-stained coveralls and leather boots, the other in jeans and laced boots. The latter was shorter, in the process of restocking wall shelves with parts. He paused, and looked up at the new office clerk typist, a head taller than himself and dressed in a white "Job interview shirt" and blue denim slacks ironed too perfectly to have been done by a young kid.

My new boss said, "Jerry, this is the foreman, Donny Ferguson. No parts or equipment comes or leaves here without Donny's inspection."

Donny and I clasped hands. I felt the squeeze of a strong hand used to heavy lifting and using precision tools. The top of his head reached my chin. He appeared fit as an ex-high college sprinter or point guard. His weathered face was hard to read until he asked where I was from. His dark eyes lit up at the word ***Ohio*** where I said I was born and raised and just completed my freshman year at Long Beach State.

He volunteered, "I was from Athens. Been here goin' on 10 years now."

I learned his home growing up was just a stone's throw from the Carney clan's abandoned New Lexington homestead. He described the oil well pump parts and mechanics of retrieving precious crude from deep underground and ocean depths. I followed him to the blacktop area behind the shop where I saw drill pipes stacked on stands and vices that secure oil well pumps for disassembling and repair. The smell of oil and solvent pervaded the air. The foreman introduced me to his working partner Rusty Logan, a stocky redhead with a pock-marked face, probably in his 40's. He was hunched over a tube casing secured in vice, gloved hands gripping a two-foot pipe cheater slipped on the handle of a heavy-duty *Oswego* monkey wrench. Standing on his toes, he took a deep breath, raised the heavy iron tool chest-high, and pressed down with all his weight and might. The sound of decoupling long-married steel pipes was like cracking leg bones. Sighing relief, Rusty removed the cheater, lowered the wrench to the pavement, and slowly recovered. Standing a hand taller than me, straightening his massive shoulders, he removed a greasy leather work glove from his right hand. I braced for an iron grip and crunching pain that came.

During the following day's lunch breaks, Ferguson said that he joined the Navy after Pearl Harbor and served as a gunner on a *Dauntless Dive Bomber* on *USS Enterprise* during the Battle of *Midway.* Recalling the images of the *Victory at Sea* documentary on TV, and aware of the incredible loss of torpedo plane crews, I was taken aback by the presence of a living hero of a historical sea battle.

Rusty impressively survived as a Marine the bloodiest and longest island invasion against the Japanese Empire-Okinawa. Over twelve thousand Americans were killed in action, more than in Iwo Jima.

But, like the rare *Midway Dive Bomber* rear gunner not shot out of the sky, ex-submariner escaping depth charge death trapped in an iron tube undersea, the 101st Airborne Division paratrooper who made night drop hours before D Day morning and fought to Berlin, and my Great Uncle Jim Carney, half-blinded by mustard gas in The Great War, I never probed combat survivors I held in the highest esteem of human race.

In the days ahead, paperwork completed, oil pump salesmen hanging out at the shop would jokingly say, "Kid, do something, even if it is wrong." Eventually, my fellow Buckeye and shop boss Donny Ferguson responded to my plight. Beginning in ride-a-longs to well sites, I'd retrieve and hand tools to him like a nurse taking orders from a surgeon at the operating table. He taught me how to measure the wear on expensive pumps with a micrometer instrument, a handheld C-shape metal frame with a moveable jaw operated by a calibrated screw gauge. I eventually came to work in jeans, and a sleeveless sweat jersey and traded my sneakers for steel-toe *Red Wing* ropers, and always wore a tin hat in the field. Told to transport a drill pipe and oil well pump to the well site with a trailer attached to a pickup, but I panicked. One look at the red flag dangling at the end of a steel tube the length of a damn telephone pole, butterflies started swarming like wasps in my guts more than waiting for Friday Night kickoff returns in Upper Arlington football stadium. The foreman pointed to the top of Signal Hill, making a circular motion representing switchbacks.

"Fuck me," I said, remembering the first California driver's license parking test, I failed.

Ferguson's succinct advice: "Kid, give yourself a wide berth."

By the end of the summer, I was a certified oil well repairman sent alone to remote well sites to diagnose, select and assemble replacements in the field or at the shop. When the fall semester started, I accepted Pacific Pump's offer to cover the weekend alone. A "down" oil well was treated as an emergency. I saved enough to buy a 1951 *Pontiac Bonneville*. Saturday and Sunday started at 6 a.m. with breakfast at *Copper Skillet* and ended at 6 p.m. with dinner at home. Maintaining B grades required long hours of secluded study in the library, consequently resulting in missed fraternity parties and friends. At the end of sophomore year, I was ready for a break. It came in a wedding invitation from the Watsons. Aunt Helen called and said that my presence was required in Columbus to serve as an usher at my cousin Jane's June wedding. My plane ticket to Columbus was reserved. Pacific Pump gave me a week off.

Jim's ordination name was "Father Vincent" Watson in the Dominican order. The church wedding of her "chaste" daughter and marriage to a "good Catholic" boy and a doctor's son was the fulfillment of the bride's mother's prayers. Wearing a formal tuxedo with tails and rented patent leather shoes pinching my toes, Aunt Helen's non-believer nephew sat in the designated family pew next to the only Black guest- Lizzy Washington, the Watson's loyal maid and keeper of children's secrets. Lizzy was wearing a bright red dress and hat. Her dark eyes glinted, reflecting the flames of thousand holy candles and ceiling lights at St. Agatha's remodeled church. Her coal-black face and glossy red lips held a broad smile, possibly enjoying the fleeting elevation in social standing. We both shared a wink and hand slap.

After the ceremony, we mingled with the Carney country folk that Aunt Helen avoided and Dr. Watson's bloodline from rural Lancaster. I avoided the stunning wives and children of doctors that tended my wounds while growing up, distinguished public officials, construction magnates, civil lawyers, judges, and real estate tycoons, many of whom were his patients and helped him challenge and defeat the AMA's war against accreditation of osteopathic doctors. I appreciated that my Uncle J.O. was somewhat an outlier in Ohio's elite medical community. His medical reputation as a surgeon was widespread and validated by the political and powerful patient caseload he never discussed, his care for working-class, public servants including mayors, school superintendents to police chiefs. As a parent, he was aloof and appeared to only interact in the adult world. My older cousin, Mono Carney, Uncle Walter's daughter, Jim's age, and Jane's bridesmaid understood the weird Watson family dynamics. She described her summer visits as a youth at Onandaga Drive when Aunt Helen roped off the living room with a red velvet rope and stanchions to protect the antiquates and undisturbed furniture; Uncle J.O.'s three-word conversations, silent dinners, and total detachment from his children.

Mono retold the time when Uncle Walter took Jim fishing in New Lexington. Jim's delight was expressed in laughter when he caught a fish. She said, "Spending a day with his kids fishing is something Jim's dad would never do."

I nodded in full agreement. My longest time with Uncle J.O other than family rides to Lancaster on Thanksgiving and South Bend to attend Notre Dame football games was lugging him and Doctors Hospital legal counsel's golf clubs 18 excruciating holes at Scioto Country Club golf course. The only physical contact was in the den when he taped my wrist *sprained* playing football in fourth-grade recess. He didn't derive any pleasure from his children and spent as little time with them as he could.

1960-1965

85 Cents Per Hour Sabbatical

With the freedom of youth, I chose friends over money.

No family to support

No expectations

After living for a few days at the Cardi's following the wedding awoke Golden Bear nostalgia of the happiest two years of my tumultuous teenage life. It was a brief era when stilted Catholic education, conformity, and boredom were trumped by self-prescribed extracurricular activity: theater arts, journalisms, unorthodox English literature, football stardom, and youthful transgressions including declared "no day-truancy," school pranks, delinquencies and borderline crimes children of privilege were fortuitously immune to legal consequences attributable to family ties and color of skin.

Only a *knucklehead* college student would walk away from a $3.50-hour blue-collar summer job without the prospect of another. The plan was to stay with Vince, pound the pavement for a job, save most of the earnings, cash in the wedding return plane ticket and hitchhike back to Long Beach in late August. I had saved $400 working weekends at *Pacific Pumps.* This would be supplemented by summer earnings and Uncle J. O.'s $500 plus tuition and books to pay for the fall semester. Thus, I'd begin the junior year with a financial cushion until I secured flexible employment on or near campus. Meanwhile, my current expenses were 3.2 beer and *Robert Burns* cigarillos. The Watsons now lived in a custom-built ranch-style home on the edge of Upper Arlington. They were not aware I was forgoing a lucrative summer job in Long Beach or for that matter, ever inquired what I did for pocket money during the school year. I faced dread and guilt over writing a letter of resignation to the best employer I ever had. I quit a job that any average college kid would walk barefoot over hot coals to secure.

I returned to my high school days' second home: a spacious third-floor refuge with separate bedrooms, bath, stereo hi-fi, and vinal library of Sinatra, Everly Brothers, Jimmie Rogers, Andy Williams, Ray Conniff, Bill Haley, Johnny Mathis, and Nat King Cole 33 rpm records. Sadly, the Cardi family dynamics had changed. William Cardi, the handsome, self-educated, erudite, and traditional Italian patriarchal husband was confronted by his wife Miriam's growing independence and rising feminism. She was still an attractive, strong-willed, college-educated Panamanian middle-aged woman who was now pursuing a Ph.D.in clinical psychology at Ohio State University, the same school where her son Vince was studying pre-law. Marital tension was evident in Mr. Cardi's chauvinistic comments. Vince worked as a summer clerk at the State highway department. The only mid-June job for students in college town came by word of mouth. A high school teammate, Bob -"church key"-McCullough, who always wore a chain necklace with a church key to pop a beer can, referred me to the upscale *Arlington Arms Restaurant* and *Motel* located on *River Road* three miles from the Cardis. Bob was a valet earning big tips on evenings and weekends. In contrast to low staff turnover of valets and all-white female waitresses, quitting was common among low-wage 85 cents an hour predominately Black busboys and dishwashers.

Wearing a clean, starched white chef's jacket, I'd start the 11 a.m. to 9 p.m. shift, harassed by homely waitresses licking their glossy red lips, picking tobacco out of yellow teeth, combing peroxide hair exposing dark roots, wiggling girdled asses as they raced toward tingling glasses retrieving tips busboys denied. My added punishment was being treated as if I was invisible, listening to belching patrons complaining over the service, taxes, mean bosses, boring jobs, and cheating spouses, while I rounded up leftovers: half-filled water glasses, empty wine bottles, overflowing ashtrays, used utensils, and dirty, wadded cloth napkins dropped on the floor.

Then I was subjected to the indignity of tumbling, balancing heavy trays on stooped shoulders, crashing into a crowded kitchen hot as a steam room, scraping half-eaten salads, entrées, melted ice cream, and fruit pie into stinking garbage cans. My final duty was being chained to the scalding dishwasher, loading, and unloading, racks of hot gleaming soylent-green-colored glasses, dinner plates, fragile coffee cups and saucers, and tons of utensils. Until I'd sneak in a free meal, escaped to a men's room cubicle or vacant suite, pounced on the king-size mattress, smoking a cigarillo, read *For Whom the Bell Tolls, Catcher in the Rye, The Plague, The Gathering Storm,* and *New York Times* articles about politicians arguing over the missile gap and spreading communism. I was *classified 1-A*, wondering if I'd survive America's next fucking war. At the shift's end, I shed a garbage-stained chef's jacket, but not the garbage smell which stayed under the skin for hours. One whiff of my scent would confirm my menial status to anyone who walked by me. It was a far cry from the pungent smell of oil and solvent that followed me home from the *Pacific Pump* shop, odors verifying I was a proud member of the skilled working class.

Notwithstanding earning poverty wages, I understood I still had many advantages from white privilege and being born in the Carney clan of geniuses and fools, winners and losers that came to the aid of kin that was thrown snake eyes by fate. I was a 20-year-old orphaned grandson who was provided succor and love at birth, education, healthcare, and freedom to play in a safe environment. Thanks to the financial assistance of a selfless physician not of his blood, and the board and room of another maternal aunt in California, I was allowed to continue my existential journey that started at a small farm and scooping coal from an abandoned strip mine. Even a few privileged high school sidekicks let it be known they envied my independence from parental set rails and expectations. Since I was not into cars, fancy clothes, or expensive excursions, I had no siblings to rival. Friends and acquaintances did not have a smidgen of knowledge of my life's history that shaped my decision to stay in Columbus earning garbage wages. Ultimately I knew it was an impractical choice, one that I could not rationalize myself. Flunking out of college was not an option any more than failing to make the St. Agatha football team as a fifth-grader or Golden Bear varsity. Football was an antidote for the self-inflicted shame of being a social outlier and not being raised in a nuclear family.

I lived with the relentless fear of being asked, "Jerry, what happened to your mom and dad?"

I imagined myself responding, "Happened? Motherfucker, you're asking me?"

The definition of an orphan is a child whose parents are dead or a child deprived of parental care and that has not been adopted. A dead parent offers survivors some closure. An absconder haunts his progeny like a ghost. I knew kids raised by single mothers, dads killed in the war, accidents, fatal illnesses, or split the scene after divorce. A child alone on a troop train or sitting on a bench at an orphanage is entitled to a face-saving fantasy to preserve the persona of social normalcy. Thankfully the first query for me ended on a train. The mystery man that branded me with my surname must've been a scoundrel. Heroes are not kept secret. I wondered what the Carneys knew about Robert Vernon Wheeler. As a child, my mother's absence was foggy, explained by kin as "illness" or "bad nerves." Symptoms I initially witnessed intermittently in Logansport at ages five and six, and again in high school. I planned to see her before I returned to Long Beach; a visit was driven by guilt and duty more than the maternal/child bond relationship precluded by years of separation and Carney clan enablers sworn to secrecy and silence that kept the ghosts at bay.

My other high school best friend, Trent Lamb dropped out of college in 1960 and enlisted in the Air Force. Had Vince been gone that summer, I'd have returned to Signal Hill repairing oil well pumps. Late July evening hitching home from Arlington Arms, I was caught in a blinding thunderstorm. Rain drenching t-shirt glued to my nipples and ribs, soaked jeans sliding off boney hips, worn sneakers squishing toes dancing on the gravel shoulder of *Scioto River Road*, all I heard were sounds of howling wind, snapping tree limbs, rattling diesel, and whining tires of 18-wheeler's splattering mud on my clothes and dreams. After changing into dry clothes, I walked into the Cardi's dining room and observed Miriam Cardi hunched over a legal-size pad, jotting notes from one of the textbooks scattered on the long oak dining table I shared countless Italian meals in high school. Enticed by a thick hardback's bold title, *Nature of Prejudice* by Gordon W. Allport, I picked up the book.

Miriam looked up and asked, "Jerry, what do you think about `prejudice`?"

Suddenly the book felt heavy in my hands. I stared at the cover, compressing my lips in silent embarrassment. I told her that the word "prejudice" rarely comes up in conversation. I described fellow students discussing their professional goals in business and public service. I related my declining interest in the *Sigma Pi* fraternity activity after the older ex-GI members moved on in their careers. I had stopped paying dues and had no plans to recruit ideal-type Caucasian gentile pledges in the fall. Pledges that reminded me of Golden Bear goody-goodies obsessed with high grades, prepping for College Boards, and building résumés for admission to elite schools.

I said, "Hell, Mim, I went to all-white schools except for the two years at St. Anthony's Catholic Boys high in Long Beach where the student body included only one Negro-the senior class president, and fellow 10th grader and fishing buddy, a Mexican kid name Emilio. At Long Beach State, one can count the number of minority students in most classes on the fingers of one hand. My friend Jim Bright won an art contest for a poster-size oil painting image of two hands shaking, one black, one white. It was published in the local newspaper and shown at a Los Angeles Art gallery."

"Do you think prejudice is natural in human nature?"

"Never thought about it," I confessed.

Mim motioned for me to give her back the book. She quickly found a passage. Drilling me with her intelligent eyes, she said, "Jerry there's prejudice and there's prejudice*:* ordinarily there is a prejudice that applies to inanimate objects, car models, or fashions. Reading aloud, she quoted America's preeminent social psychologist *Gordon W. Allport, The Nature of Prejudice*: "prejudice as an aversive or hostile attitude toward a person who belongs to a group, simply because he belongs to that group, and is therefore presumed to have the objectionable qualities ascribed to the group.

In the evenings that followed, I read, Allport's chapter, *Rejection of Out-Groups:* "When the excitement grows high during a lynching in the south, it is taken for granted that there will be looting and destruction of Negro homes and businesses. Not infrequently furniture from Negro homes is used as firewood for burning the victim's body." The author's description of mob lynching, mutilation, and murder reached its zenith in the Reconstruction Period after the Civil war but continued in the 1940s and into my life. But incidents of violence, rape, and discrimination against Negroes by the Klu Klux Klan and advocates of white supremacy that prevailed in the south were not headlined news in California and Ohio anymore that my contemporaries understood the history of the Federal Government's broken Indian treaties, Calvary massacres and Trail of Tears documented in Grant Foreman's book, *Indian Removal: The Emigration of the Five Civilized Tribes of Indians.* Unlike Hollywood images of native Americans in western movies like *Little Big Horn, Buffalo Bill, and The Searchers,* a white teen's tourist stop remembrance of Route 66 Trading Post, a kid following convicts marching lockstep into a prison chapel, playing with Negroes and Appalachian poor at Roosevelt Recreation Park at Eastside Columbus, childhood years bonding with the Watson' maid, and later working side-by-side with Negro busboys and dishwashers trapped in a culture of poverty. But, thanks to Carney family ties and fate's roll of the dice, I understood on a personal level, each shitty job from hurling newspapers at porches of mansions in drenching rain and blinding snowstorm, pumping an ice cream cart uphill, scrapping rich patron's leftovers into garbage cans, was a privileged white boy's "temporary" experience in life's adventurous journey. Indeed, under Miriam Cardi's wise tutelage, I experienced an epiphany over America's Out-Groups' inescapable suffering and abuse buried in the news and ignored in high school civics and American history.

In mid-August, I experienced headaches that slammed me on the mat like a sumo wrestler. I couldn't breathe through my nose. Nasal passages expelled green slime. Vince's mom insisted I see a doctor when applying a hot pad on my forehead and aspirin offered no relief. Aunt Helen referred me to her husband's colleague Dr. Ralph Licklider, the ENT specialist that removed my tonsils. He examined and x-rayed my nose which was broken repeatedly during my short lifetime by playing high school fullback before face bars were considered standard headgear. The diagnosis was *chronic Sinusitis*. I was admitted to Doctors Hospital where Dr. Licklider performed a *submucous resection* procedure which removed the bony and smashed cartilage that obstructed the nasal passages and caused sinus headaches. As I was wheeled on a gurney to the OR, I passed my college benefactor, Dr. Watson himself, wearing surgical scrubs and going in the opposite direction.

I awoke on my side in excruciating pain of sharp spears jamming up my nasal passages into my brain, puking bloody goo in a bedpan pressed against my chin by black hands reaching over the gurney rails entrapping my somnolent body. Eyes adjusting to the light, a black face appears; Black as the Great Satchmo's I met with Cardis at Columbus Hilton. I heard a deep scratchy voice say in a compassionate tone, "It's the cotton packing, kid." He wore a white orderly uniform splattered in human blood. He carefully walked to the toilet, flushed my regurgitated sinus pus, returned, and wiped goo away from my face as I pleaded for pain relief. He said, "Nurse is on her way," then made his adieu.

The next image was a stocky, gray-haired white woman wearing a tiny white nurse's cap, white dress, name tag, white hose, and white shoes. Her pale hands fumbled with a tube of saline drip and an I-V needle inserted into the vein back of my hand. Next, she retrieved a syringe, gestured for me to turn on my side, jabbed the needle in my butt, and whispered, "Son, this will help the pain."

No visitors from *next of kin*, ha. No big fucking deal. Aunt Helen never visited when I had a tonsillectomy, called, or wrote during my California exile, or away at college. Jane, whom I once adored like a big sister, never spoke of my exile, or return, or asked me shit. "How is Geraldine Roberta doing," like she did when we were kids. We'd become estranged. Her brother Jim, who was a priest, and a caring soldier of Christ, didn't reach out either. I waited anxiously for the doctor to remove the nasal packing impacting my brain when Vince and his mom made a surprise visit, bearing gifts: a book and flowers. Miriam noticed the wall niche was missing a TV. The nurse told her that a TV costs patients extra. Miriam said in disbelief, "The Watsons didn't spring for a TV."

But I gave it no thought. My aunt would have considered a TV a luxury in a private hospital room for her kids. Unlike other privileged Arlington parents that give their high school and college-bound offspring an exorbitant allowance, a car, brand clothes, a private telephone, TV, and stereo in the bedroom or dorm, The Watsons intentionally avoided giving the message of "spoiling" their children materially or using the *rod* to correct behavior. They rejected conspicuous consumption and were the last to purchase a TV. Notwithstanding her lack of maternal affection toward any of us, I truly believed Aunt Helen would not visit Jane or Jim in my circumstance.

Miriam slid the hospital cart table away to make space for a chair at my bedside. Her long hair framed her ageless, tanned face. She must have come directly from a class at OSU because she wore a casual buttoned summer blouse, jeans, and sandals. Normally at home when Mr. Cardi was in town, she wore a dress and street shoes. Scanning the dark circles forming under my eyes, she leaned over and gently touched my pale face in an act of maternal affection that I'd only experienced as a child under Grandma Carney's care. She asked, "Are you in pain?"

Teary-eyed, I was too embarrassed to answer her in my muffled nasal twang. I was in pain, but I had to wait another damned hour for the next scheduled morphine shot. Vince, handsome and dressed sharp as usual gave me a prankster grin, planted flowers on the windowsill he probably stole at a gravesite, stood by his mom, and made jokes about me being assigned the oldest nurses at *Doctors Hospital.* He whispered, "Buffoon, you looked like shit."

The public address system announced, "Visiting Hours are over." Miriam hugged me and introduced me to Eric Fromm's *Escape from Freedom.*

The next time the nurse jabbed a needle in my vein, a TV was in the room. I watched *Gunsmoke, Red Skelton,* and *Have Gun Will Travel* and western re-runs at night and read during the day. I was aghast at the rope-size string of bloody cotton that Dr. Licklider slowly pulled out of my nasal cavity. The wondrous physical relief of pressure in my cheeks and forehead that arrived as he pulled reminded me of a magician pulling endless handkerchiefs from his pocket. The next day, Vince and Miriam were joined by other visitors from the Golden Bear days. One was Joe Arnold, an underclassman football linesman. He was accompanied by his blonde girlfriend, Carol Swanson.

In no physical shape to hitchhike back to Long Beach and pocket the airfare as planned, I finished reading *Escape from Freedom* flying above the clouds in *TWA Constellation* the day after hospital discharge. Reflecting on Fromm's thesis: *Destructiveness is the outcome of unlived life*; how man's fear of his loneliness drives him to surrender *his freedom by transforming himself into a small cog in the machine, well-fed, and well clothed, yet not a free man but an automaton.* This dynamically spawned fascism led by Hitler. The power of one and self-identity succumb to the group leader's decrees, dress code, and symbols. A thesis-driven home after I completed William L. Shier's masterpiece, *The Rise and Fall of the Third Reich: A History of Nazi Germany.*

I arrived at LAX, weaker and thinner, a week before fall registration. I had a $600 check issued from the Dr, James O. Watson's medical office, $310 cash savings from scrapping garbage, and $400 in Long Beach bank saved from Pacific Oil Well repair days. The 1950 *Pontiac Bonneville* covered in dust, sand, and bird shit was parked where I left it in front of the Paskell's residence. The battery was dead and the tires were begging for air. The interior reeked of salt air and dried *Jim Beam* puke. Upon assessing my net worth, jobless state, the cost of the tune-up, battery, gas, and insurance reality stepped in. I managed high school, the first college year without wheels. I had a bike, two feet, and a thumb. The campus was geographically close. Students started riding bikes. No shame in that.

My cousin Bill was a senior at *Milligan High*, still driving a stagecoach chased by white actors playing renegade Indians at *Knotts Berry Farm,* Aunt Marg still counting other people's money and Marjean sleeping over with friends more than at home to keep out of her mad father's range. My first call was to a junkyard dealer. He paid me $18 to tow away a rusted green steel albatross. Ernie said I was nuts.

New Horizons

The ex-GI frat brother that tutored me through dumb-bell English composition moved on to a teaching position in Torrance; replaced by younger clean-cut aspiring coaches, teachers, entrepreneurs, bankers, and accountants-fraternity bros. My best friends Nick Gordon and Jim Bright rented a cheap apartment in Sunset Beach, a growing community of artists, and students 10 miles down Pacific Coast Highway. Also hanging out there were small groups of tanned, long hair, surfboard fanatics emulating local pioneers like Dick Brewer and Dewey Webber observed sanding and waxing fiberglass boards on makeshift construction horse benches in dilapidated garages, waiting for the next optimal offshore wind and barrel wave ride to fame. Nick and Jim were avid surfers under no pressure to carry a full load of class to avoid the draft because they completed six months of active Army reservist training straight after high school and then committed to monthly weekend duty; my plan upon college graduation. My frat buddy Jack Wilkerson continued to commute from home driving a small used English car that sounded like a sewing machine on wheels.

I declared social science a major, but I regretted signing up for a business law course taught by a part-time ambulance chaser who took pleasure in drilling shy students on cases we never read or were interested in. I gladly left behind boring *Stare decisis*, welcoming delving into the *Civil War* History and *Reconstruction* taught by an enthusiastic protégé of Bruce Catton that keep our undivided attention from Gettysburg to Appomattox, atrocities of lynching and Jim Crow. I listened to a WWII combat vet and Political Science professor who abhorred sleeping in foxholes, olive drab color clothes, always graded blue book exams on the curve, and maintained, “Money is the blood of Politics.” The smallest class-*Soviet culture*- was taught by a Russian immigrant that focused on the pathos of Russian literature, the oppression of the czar, and Stalin’s genocide.

Doctor Watson’s largesse carried me through Christmas. But I needed to pocket money for smoked fish and beer picnics at the beach, and *Joe Jost’s* 25-cent schooners of draft beer served with pretzels and pickled eggs. Most importantly, I needed a job that would allow me to move out on my own. Aunt Marg’s Catholic background appeared to inhibit her from divorcing her volatile husband whose slumped shoulders and posture conveyed the image of failure and defeat as Willy Loman in *Death of a Salesman.* Marjean was a good kid and an excellent student. But at home, she lived in fear, like a docile dog waiting for its master’s next kick. When will she escape?

I was walking out the main exit of the administration building, down to a few hundred dollars after paying spring tuition when I ran into a guy my age unloading a dolly and boxes from a gunmetal gray pickup with **LONG BEACH STATE COLLEGE** painted in bolt print on the door. He was dressed like a student rather than a maintenance staff recognized in *Dickie* work clothes, clodhoppers, and photo I.D. Many were middle age Navy retirees responsible for tending to the landscape, custodial repair, seen driving on the wide campus walkways and delivering equipment, office supplies, and campus departmental mail. Holding the door for him to pass with the dolly, it dawned on me that the other door was the entrance to the college President's office complex. We were immediately confronted by a tall brunette beauty with laser-dark eyes that froze us in place, as she slowly rose behind a mammoth desk. I fought back stargazing at Helen of Troy as she led us to a small supply room next to the college President's office. Holding the supply room's door for the dolly to pass, her body language communicated the deep pleasure of knowing the power of her sexual presence over two horny young dudes stumbling into her space. The dolly-pushing kid said the rumor was that the college President's secretary was an ex-Miss Universe which from all appearances must have been true. On her desk was a photograph of her smiling and leaning down with her arms around the shoulders of a handsome man in his late 20's sitting up in a wheelchair. He was wearing a team jersey that has the image of a wheelchair with wings and words, ***Flying Wheels*** I recognized as the name of the national champion basketball team made up of paraplegic war veterans. The photograph took me back to images of wheelchair-bound men wrapped in Army blankets assisted by nurses on the sidelines of the *Chillicothe Veterans Hospital* football stadium. And the legless veteran whose propped-up torso I saw teetering on a plywood platform with rollerblade wheels he maneuvered by fingerless gloved hands-on busy Columbus street corners; casualties of war holding up what was left of human dignity in a tin cup, selling pencils to passersby, eyes askance.

The dolly-pushing student said the school paid student assistants $1.25 an hour. If interested, he'd drive me to the employment office at the vocational education building at the lower campus.

He introduced me to a burr-headed leather face man sitting erect behind a government-issued metal desk decorated with a miniature Iwo Jima sculpture and Marine Corps flag. He said, "Kid, before you bother to complete the application, let you know this job involves more than cruising the campus in a State-issued pickup looking for pretty coeds. You'll be policing student garbage and litter, plunging toilets stopped up with *Kotex* and condoms, fire-hosing sand-stormed tennis courts, painting over lewd drawings in heads. Know what a **head** is, kid?"

"Yes sir, I live with an uncle who was an ex-submariner in the Pacific."

He nodded approval, saying, "Job pays $1.25 an hour for working 20 maximum flexible hours a week, paid once a goddam month, no raises, no paid sick leave or holidays. Still interested?"

"Yes, Sir."

"Got a valid *California* driver's license?"

I retrieved it from my wallet and watched him examine it.

"Ok, Gerald, you wouldn't believe the students who don't. You sign in here for a vehicle and the necessary tools. Driving rules are strict. No speeding, giving buddies rides to class. You wreck a pickup, hit a pedestrian, you're terminated, drive off campus unauthorized, you're terminated."

My starting assignment was responding to a plumbing problem on the second floor of the women's dorm. I signed out for a 1950 Chevy pickup, and retrieved a bucket, mop, soap, cleaning rags, sponges, rubber gloves, and plunger. Thought this was a good opportunity to explore an alternative living arrangement. Procedure dictated that a female resident supervisor escort a male repairman because girls may be running around half-dressed. Wearing a bathrobe, flip-flops, and hair in curlers, my escort met me at the top of the stairs. She formed a megaphone with her hands and screamed, "Male on Floor! Male on Floor!"

The sound of opening and closing doors and giggling female voices echoed down the hall. We proceeded to a communal bathroom. Suddenly its double doors burst open, and girls flew out, flashing panties and bras, snapping towels, yelling, "Man on Floor, Man on Floor," in affirmation. The escort rolled her eyes, and waved me inside, announcing, "All clear."

Our wide eyes in sync scanned the aftermath of a hurricane: flooded tile floor, six porcelain sinks and counters strewn with open cosmetic jars, squeezed toothpaste tubes, toothbrushes, lipsticks, makeup-smeared tissues, and cotton balls, nail polish, hairdryers, curlers, combs and brushes shedding hair, soapy towels heaped in communal shower leaking faucets, Reacting to the dorm resident supervisor holding her nose and pointing at a private stall, I attacked an overflowing commode with a plunger and rubber gloves. Without apology, exiting the scene the female voice said, "Welcome to the women's world."

Next work assignment, I found myself standing in a wind tunnel struggling to hold a heavy brass nozzle attached to a heavy fire hose turned on full pressure, blowing me backward as water blasted the tennis court after a sandstorm. I quickly understood why firemen wear raincoats, rubber boots, and special plastic helmets. I cleared $80 a month and often partnered with full-time support staff that appreciated my mechanical skills beyond responding to stopped-up toilets. The best runs were delivering interdepartmental and U.S. Mail. The campus post office was across the street from the vocational education academic and support staff administration building that scheduled student work assignments, documented work hours, and issued payroll checks. As days went on, the Paskell's welcome mat became frayed. Weekends, Thanksgiving, and Christmas holidays were spent at the homes of Nick Gordon, Jack Wilkerson, and Jim Bright.

Christmas break, I leaped from my Stearns Street nest, landing at a vacant school dorm rent-free until the hallways teemed with returning students. I responded to the college's message board post: "Roommate Wanted, Seal Beach apartment."

The address was a 1-bedroom garage apartment adjacent to an alley a block from the beach. The entrance was a wooden stairway built like a fire escape on the side of the frame structure. A swarthy skin burly man with a patch of dark hair, black pearl eyes, bushy eyebrows, *the Bluto* face of *Popeye the Sailorman*, shimmering pectorals, nipples pressing against a faded athletic jersey with cutoff sleeves and holes matching sweatpants cut-offs that revealed the hairy arms and legs of a bodybuilder. A stump neck atop massive shoulders and a tall frame almost filled the door space.

I entered a one-bedroom linoleum floor apartment architecturally the size of a two-car garage below it. The living room furniture started at the door's immediate right: a wine-stained, lime-colored three-cushioned Hide-Away day bed, a gray tweed-tattered fabric armchair, and a black metal floor lamp. The west side offered natural light from double-pane windows with a second-story view of telephone poles drooping wire, swaying fronts of tall palm trees, steeple shape churches, and flat roofs of residential dwellings, apartments, and businesses catering to the small beach community known for its art galleries, small cinema featuring foreign movies, public fishing pier, clean beach, and ocean horizon. Snug against the oyster-color wall, adding to humble décor was a walnut coffee table overburdened with a telephone book, rotary telephone, a glass ashtray, sport, and men's health magazines, and a transistor radio. Directly across the sofa was a door to the bedroom and adjacent bathroom. The bedroom contained a double bed sinking mattress covered by a bedspread. The head of the bed was on the west window side of the room. The foot, a short hop to a truck stop motel: small sink, shaving mirror, towel hooks, a commode, and curtain enclosed standing-room shower and. Drain. The kitchen was on the alley side. The table was befitting a teardrop camper plus two flea market chairs. A small gas stove, fridge, sink, and dish shelves were surprisingly clean and organized.

My roommate's name was Dave Hickok, a senior anal compulsive recreation Gene attending on the GI Bill following four years driving heavy equipment in the Navy Seabees. The $50 monthly rent included utilities and a $25 fifty-fifty share for a phone. We took weekly turns by sleeping in a double bed and a sagging sofa bed. On Sundays, we washed bedding at the local laundromat. Neither of us had a girlfriend but agreed should the opportunity for a female guest arise, not to enter the apartment if the stairs light was on. Dave gave me a ride to school in a 53 *Ford* coupe Flathead v8 he compulsively washed and waxed and covered with a tarp to protect the metal, paint, and chrome from salt air corrosion. I caught a ride home in Nick's or Jim's VW bug on their way to their Sunset beach apartment or thumbed.

During the daily assignment of picking up incoming U.S. mail at the lower campus office before making deliveries to departments, I formed a friendship with Pete Haggerty, a mail clerk whom I'd previously met at the *Seal Beach Laundromat* he co-owned with his dad. Pete was gathering coins from the machines and soap dispensaries. Pete's forehead and skull were seriously disfigured: dents and ear-to-ear scars arced over his shaved head. He jokily said, "My skull and brain are held together by a metal plate." He proudly wore a University of Nebraska Phi Beta Kappa college ring. He told me he was a Captain and tank commander during the *Battle of the Bulge* when his *Sherman Tank* took a direct hit from a German 88. Medics found him half-frozen in the snow next to the turret blown off the tank's hull, and the bodies of his crew. Medics said the subzero temperature coagulated his blood, saving him from bleeding to death. Pete was the most seriously wounded combat vet employed at LBSC. His father drove him to work and everywhere else. My speech professor had one lung lost at Anzio. A math teacher wore a prosthetic leg after stepping on a land mine on D-Day.

Stage Fright

I awoke to the sound of Seal Beach's 6 a.m. foghorn, scooped a bowl of *Cheerios,* and rode to campus with my roommate-Popeye's nemesis-a compulsive health nut and introvert. I coordinated my work schedule, classes, and studying in the library until dusk. After I'd hitch, hike, or catch a ride home with Nick or Jim, returning to their Sunset Beach abode. For backup transportation, I purchased a used three-speed English bike that made for a hard ride pedaling uphill on Pacific Coast Highway to campus but coasting home was fine.

The most anxious academic challenge in the spring semester was participating in a one-act play project written and directed by a theater arts graduate student completing his master's thesis requirement. The play was presented on stage to a live audience. I was one of three undergraduate student actors playing siblings: two brothers and a sister struggling to keep the family farm during a severe drought. The scene was in a kitchen setting and involved a family argument over how to prevent the bank from foreclosing the property. I was the youngest sibling, sitting at the table with my pretend sister mulling over our rehearsed lines that followed our older, extremely obese, and irritated brother standing across the table, before continuing his lines. The script called for him to yank out a Captain's chair in anger and assume his rightful place at the table as head of the family. As rehearsed, he swiftly pulled out the chair and squatted to sit. Suddenly, in the microsecond, his barrel-size buttocks wiggled in the tight armchair's seat, and the wooden legs snapped, crashing into pieces under the massive body landing on the stage floor in a beam of bright lights to sounds of human cries of the audience reacting to the unscripted sight of Humpty Dumpty in overalls sprawled on stage, struggling to rise, propping his elbows and knees, grasping the top of the table to balance standing suspending time and motion.

I wasn't confident he'd recuperate.

The fictional sister stared wide-eyed at me then at the fallen sibling. My instinctual reaction was to improvise off-script and break the graveyard silence in the theater. But Humpy Dumpty gritting his teeth, leaned down to the floor, retrieved splintered legs of the wooden chair, then slowly stood erect, facing the audience, raised the broken pieces in the air, and in a raised voice said, "Damn, *S & H Green stamps,"* which brought the house down. Like most of the public, the audience had redeemed hundreds of *S & H Green stamps* for cash or household items including toys, dishware, and furniture offered in the *Sperry & Hutchinson Company* catalog. Customers received stamps at checkout stations at gas stations, department stores, and other retailers that used these trading stamps as incentives for shoppers.

In June, my roommate graduated and left town. I canceled the phone and rode my bike to work and completed one class, *Conservation of Natural Resources* taught by a natural scientist from Hyannis Massachusetts, rumored to have sailed with JFK at Cape Cod. My working hours doubled in the summer months. State-issued monthly checks supported a Spartan lifestyle. The plan was to save enough to share a three-bedroom Sunset Beach with Nick and Jim in the fall when rentals are significantly reduced. I worked as the campus utility man and learned how capitalism was destroying the earth. No car to pursue dating girls or distracting TV, I purchased a dozen used paperbacks at *Acres of Books* in Long Beach: Henry Fielding, Hemingway, Hugo, Somerset Maugham, Hess, Camus, and a small treasure, *The Execution of Private Slovik* by William Bradford Huie, I found in the dusty stacks.

Private Eddie Slovik was the only American soldier executed for desertion in WWII and since the Civil War. He was a troubled youth: in and out of jail and called a *loser;* just as he was turning around his life, married, working in a factory, and discharged on parole, he was inducted into the army in early 1944. He hated rifles and considered living in the army comparable to life in jail. Slovik was one of 12 replacements sent to Europe to join Company G, 109th Regiment, 28th Division. On August 25, they were introduced to the carnage of war in a truck ride passing the aftermath of the fighting in the *Falaise Gap*. When they arrived late at Elbeuf, France, the city was under shellfire. Ordered to dig in outside the city, Slovik found himself *nervous and trembling* in a foxhole during a barrage. After it was lifted, he and another replacement were lost in the confusion. The next day his new unit did not realize Slovik and his replacement buddy were missing when it moved out to another mission. The city was turned over to a Canadian military police outfit. For the next six weeks, the pair of GIs befriended the Canadians driving trucks, cooking, and guarding German prisoners. Believing a jail sentence would be the worst punishment, Slovik turned himself into Army authorities, confessing in writing that he was a deserter and demanded a court-martial:

I told my commanding officer I, Pvt. Eddie D. Slovik, 36896415, confesses to the desertion of the United States Army. At the time of my desertion, we were in Albuff in France. I came to Albuff as a replacement. They were shelling the town and we were told to dig in for the night. The following morning, they were shelling us again. I was so scared, nervous, and trembling, that at the time the other replacements moved out, I couldn't move. I stayed there in my foxhole till it was quiet and I was able to move. I then walked into town. Not seeing any of our troops, I stayed overnight at a French hospital. The next morning I turned myself over to the Canadian Provost Corps. After being with them for six weeks, I was turned over to my story. I said that if I had to go out there again I'd run away. He said there was nothing he could do for me so I ran away again AND I'LL RUN AWAY AGAIN IF I HAVE TO GO OUT THERE.

— Signed Pvt. Eddie D. Slovik A.S.N. 36896415 ....November 11, 1944.

In less than two hours of deliberation, the nine-officer court-martial panel passed a unanimous sentence. *Under the Articles of War*, the penalty for wartime desertion was death; to be *shot to death with a musketry*. Given thousands of GI casualties and record surrenders to Nazis in the Ardennes Forest in The *Battle of the Bulge*, in the U.S. Army's perspective, *it was a bad time for mercy* for a recalcitrant soldier who refused to fight. His final appeal, on December 23, 1944, was made-to Gen. Dwight D. Eisenhower, the Supreme Allied Commander. Eisenhower upheld the death sentence. On

January 31, 1945, a 12-man firing squad in Sainte-Marie-aux-Mines, France, shot Private Slovik. He told a guard that he was being executed, "for bread I stole when I was twelve years old." Slovik was buried in an unmarked grave in a special section of the Oise-Aisne American Cemetery in France alongside 94 soldiers executed for rape or murder. I pondered, was capital punishment for desertion, justice? Of millions conscripted in two world wars, Slovik was the only soldier executed for desertion. Even Lincoln reprieved many Union deserters. I reflected on the case against mercy for Slovik. Visualizing tens of thousands of crosses and Jewess stars in national cemeteries and France alone, the wheelchair-bound and walking wounded straggling out of Chillicothe VA hospital to watch healthy young boys crack their skulls and bones, or bruise bodies in their rights of passage in America's most violent physical contact sport. How fans cheer for the injured players carried or assisted off the field! But the veterans wrapped in Army blankets hunched in wheelchairs on the sidelines and others sitting shoulder-to-shoulder in tiers of wooden bleachers, chain-smoking, and the legless casualty of war selling pencils in a tin cup on Columbus Street corners were met with silence. How would they or my friend Pete Haggerty whose hopes and autonomy were permanently shattered by a Nazi 88 anti-tank round in *The Battle of The Bulge*, or my Great Uncle Jim gasping for breath and nearly blinded by mustard gas in WWI, or the paraplegic vets *Flying Wheels*' judge Slovik's appeal for mercy?

But of all the words and images of suffering of victims of war, my most visceral reaction was to a black and white photo published in *Life WWII Photographs.* Seared upon humanity's collective was an image of a blindfolded young man my cousin Jim's age, kneeling over a freshly dug grave, hands tied behind his back, wearing khaki shorts and combat boots. He was identified as an Australian pilot prisoner-of-war. Hovering over him was a Japanese officer in fancy leather boots and flaring riding trousers, brandishing a samurai sword, its blade at the arc of downward thrust of decapitation: a war crime frozen in time. The first time I saw the image, I held my breath. I pondered the young pilot's final thoughts, visualizing the pain of tempered steel slicing through his neck's flesh and bone, his headless body tumbling into the grave. I shuddered, the severance of the human bond between the captor and prisoner never more evident than at that moment.

Riptide

Final senior grades posted. With my Sunset Beach roommates visiting their parents, I donned baggy trunks, retrieved bodysurfing fins, stepped outside, and responded to the rhythmic sound of lapping waves' beckoning call restless June afternoon. I inhaled the salty air, expanded my lungs, and gazed at warships crossing the blue-green ocean horizon in the cobalt sky. Slowly I exhaled, expelling rote memories of academic texts, then raced barefoot across hot sand until my toes were rescued by foaming surf. I dove backward in the tallest wave, struggling underwater to slip each foot into a web-shaped hard rubber boot. Boots lead fins, suddenly my body spinning topsy-turvy in strong clashing currents, trapped in turbulent water. Water churning driftwood, cone shells, rope ties, plastic bottles, fishing tackle, fossils, and bones. Instinctively, I kicked my flippers like a desperate mermaid springing off an underwater dune toward the light, bobbing head surfacing in rolling swells. Lungs bursting, I spat salty water, limbs useless as a puppet dangling in tangled strings. *Deja Fucking Vu!* I mused, executing the last flipper kick before being sucked into the ocean grave, eyes squinting at the blurry image of a man carrying a surfboard. I reached for the sky, screaming, "Help, I can't swim! I can't swim!"

I lay on my side too weak to rise, expectorating bitter water, sand, and plankton, my lungs gasping for air. I looked up at the tall off-duty lifeguard god who paddled the longboard through swelling waves against the tide to save me as if I were a toddler drowning in the baby pool. So embarrassed, I apologized for my ineptness and lack of swimming skills. According to Nick and Jim, both avid surfers, Certified Ocean LifeGuards were highly paid civil servants that must pass rigorous training, running in the sand then swimming a mile towing equal weight in the turbulent sea; ninety-five percent of applicants fail the test. I, without a doubt, remained solidly in the ninety-five percent thanks to the god-like five percent.

A week before graduation, I received a letter from Aunt Helen instructing me to pick her up at LAX the day before commencement. She arranged to stay with her younger sister-my Aunt Marg and crazy Ernie at their house in Huntington Beach. Accustomed to luxury Buick comfort, she laughed like a child experiencing adventurer's delight when I adjusted Nick's borrowed VW beetle's shoulder strap for a dodgem-car freeway ride of her life. I had never heard her laugh louder in her life.

"Free, White And Twenty-One"

In June 1962, I enlisted in the California National Guard that deferred reporting to basic training at FT Ord until September. The plan was to follow my spirit trail before commencing six months of active Army Reserve service. Thanks to Congress's reaction to Sputnik, I supplemented the student job income with a $2000 low-interest loan grant in spring. By some bureaucratic snafu, a government official must have classified *Social Science* in the same category as rocket science. Ha.

During this period LBSC campus was oblivious to the civil rights movement in the South but I was not. I quit and renounced the exclusive white fraternity that stood for white privilege. I was keenly aware of the advantages I received just by being born to the group in power and the effect of life's exigencies on the poor and elite and the in-between. I followed news reports on the "Freedom Riders" bus firebombed in Anniston, Alabama by Ku Klux Klan. Other civil rights protestors were beaten by cops, arrested, and jailed in Jackson, Mississippi in 1961. Horrific. Within a few months, police arrested more than four hundred Freedom Riders. Driven by personal outrage over police and white citizen violence against civil rights protestors, I found myself standing outside the Long Beach Naval Base entrance, counting on a ride South with sailors on leave.

With $50 in my socks, and dressed like a college student, I brought only the essentials: a *Texaco Road Map* tucked in my hip pocket, a backpack filled with a canteen of water, a change of clothes, and toiletries. I was soon picked up by two young white men in civvies driving a 1960 Chevrolet Impala with Georgia license plates. Told them I wanted to visit New Orleans before reporting to Army boot camp. If I said, Jackson Mississippi-not known as a tourist town- for white Yankees, it would have raised suspicion. Speaking in strong southern accents, they reminisced about past wild high school days in Atlanta and their many female conquests on shore leave. The discharged U.S. Navy 2d class Boatswain mates traded off driving. They contrasted civilian freedom to the "Bullshit disciplined Navy life slaving on tin cans and cruisers," and expressed strong resentment of "taking orders from `nigger` petty officers." Their racist rants increased the closer they got to home. The racist slurs I rarely heard in the circles I ran with were applied to all blacks I'd ever known: playmates at Roosevelt Recreation Park in Columbus Eastside, the few Black LBSC students and employees, and black football player teammates. Even though blacks were invisible in southeast rural Ohio coal country and working-class Grandview, they floated around the background in privileged Upper Arlington suburb life, Stay-at-home housewives referred to their Black maids as "colored" to family and friends, but not to their faces. The first Black face I saw in the flesh was a porter at the Logansport train station when I was five; the first "Whites Only" drinking fountain I spotted in Texas in 1954.

From a young age, I was cognizant of segregated housing and education in the North and discrimination in public accommodation in the South. Every child was; this was how indoctrination happened. Very few people questioned the racist messages; most adopted the messages or at least tolerated them. Racism was not questioned publicly, at least not on a larger scale until the 60s. In June 1962, images of casualties of white supremacy were headline news.

Seething in the back seat of the Impala, I wanted to call the racist ex-swabbies out: Stop! Let me Off! But that would defeat my mission to better understand the depth of the civil rights struggle and associated violence erupting in the South.

It was twilight when the ex-sailors dropped me off at a carryout in Jackson. The cashier gave me directions to the *Greyhound* bus station. I was walking alone downhill when I saw a tall figure slowly ascending in the distance. He was assisted by a cane. The moment we recognized the color of our respective faces, the old man paused, bowed his head, using his cane to balance, side-stepping off the sidewalk into weeds in a vacant lot. He stood with his head bowed while I passed in discomfort. The image of a gray-haired, wrinkled black face senior hobbling uphill, giving a healthy young white guy easily negotiating downward stride, a fucking, *by your leave*, dropped my jaw and made me feel ashamed.

The Jackson *Greyhound* bus station was teeming with travelers. But in contrast to the integrated facilities in Columbus, Ohio, and Albuquerque I experienced as a child with Grandma Carney and later as a runaway, the Jackson *Greyhound* bus station had segregated restaurants. I followed black families into the *Colored Only* restaurant and sat at the counter occupied by only dark skin patrons. I asked a black waitress for a Coke. She raised her brows, nodding toward the other side of the bus station reserved for *Whites Only*. My rationale for not joining civil rights advocates was time constraints, ignorance, impatience, and my independent nature. I was, after all, a college graduate flush with dough and only given a 90-day pass before completing one's patriotic duty. I couldn't help but feel guilty over the gifts of fate a passed-around orphan of the Carney clan received, a clan that imparted values of humility, compassion, empathy, independence, and self-reliance.

Sipping the ice-cold *Coke* sizzling in the glass, it reminded me of drinking bottled Coke filled with peanuts with kids in Anna, Ohio while visiting Grandma Carney's WWI disabled brother. Suddenly, I felt strong hands pulling the back of the collar, the shirt's top button pinching my throat, my body falling backward off the stool, then being dragged on the hard floor out the door. A strange voice spat in my face, "fuckin Yankee shithead needs a lesson of southern hospitality."

Looking up from the sidewalk, I saw two brutes out of Hollywood central casting auditioning for *Ku Klux Klan* roles approaching me: rumbled full brim gray hats low on the forehead, accentuating dark hateful eyes, face weathered by sun and draught, snarled open mouth missing teeth, chin dripping chew on oil-stained khaki shirts, faded jeans and cracked work boots. They stood me up in front of two uniformed cops waiting by a police car. Biting my lip, I was unable to resist, I felt the cop's hands patting me down and pressing my head downward, forcing me to stoop down and slide into the cruiser's rear seat. A bystander tossed my backpack to the cop entering the passenger's side front seat. He took a quick inventory, then threw it on the floorboard, leaving no trace I was ever at the bus station.

We passed storefronts, thrift shops, gas stations, cheap hotels, bars, and nightclubs flashing neon signs, streetlights blinking alive, jaywalkers and hookers darting in and out of shadows, Army and Navy recruiting posters, street watchers drinking booze concealed in paper bags, smoking pipes and cigarettes or spitting chewing tobacco. All they had in common was moon glare reflecting on white faces and lazy eyes gazing at passing cars, vintage pickups, and rattling 18-wheelers escaping gridlock under looming clouds and dying history. It should have been a short ride to jail. My anxiety increased as we crossed through the town and continued along until there were fewer signs of civilization, and the sound of tires kicking gravel indicated the police car changed from a divided highway to a widening narrow unimproved road. The only light was the beams of the cruiser's headlights shining into the darkness.

My heart raced as the driver reduced the speed and slowed to a stop.

He rolled down the window and lit a cigarette, smoke drifting in my face, stinging my eyes. I wiped away sweat blurring my eyes with my shirt sleeve.

My throat was dry.

The canteen was in the pack.

The pack was on the front floorboard.

I listened to the engine idling in concert with my heavy breathing and bullfrogs croaking my last rights. Fuck me. My body stiffened. My night vision revealed a landscape of foreboding doom. The cruiser was parked off-road but not in the typical forest of pine, maple, oak, aspen, and sycamores seen in California and Ohio national forests. This was a land of giant ubiquitous pyramidal shape bald cypress trees with knotted roots, trunks six feet or more in circumference, drooping branches draping moss and leaves canopying ink color swamp. It was a swamp crawling with venomous snakes and alligators, and bones of slaves buried by overseers of antebellum plantations. Equally unnerving to a naïve ***free, white and 21*** wanderlust youth far from friendly roots was recalling news reports of "unsolved swamp murders," "lynching" and beatings of civil rights workers in the heart of the Confederacy.

Suddenly, the two-way radio crackled, static jolting my brain numbed by fear. My captors wore police uniforms, not Klan robes or dressed like the slovenly dolts that dragged protestors out of the *colored-only* restaurant. I reflected on previous encounters with cops, starting when an Upper Arlington rookie rousted me out of a tree fort, I was constructing in a vacant lot adjacent to the. I was eight. He escorted me to our kitchen door as if I were a goddamn criminal, saying he was responding to a neighbor's complaint, "A boy was destroying private property."

Aunt Helen grabbed my hand, stood on her toes, leaned in the young rookie's face, and said in a rebuking tone, "Officer, this ***child*** lives here. For your information, this is Dr. James O. Watson's home. Dr. Watson is your Chief's doctor..." Looking over his shoulder at the complainant's house, she said, "My God, such matters are best resolved between neighbors, not in arresting children."

The young cop tipped his cap and retreated. He said, "Yes, mam."

The second encounter was in high school's junior year: Stop & Frisk by the local constabulary. Two uniforms jumped out of their respective cruisers blocking my path to walking home. They asked why I was there. Said I was walking home from my girlfriend's which was true. They examined my Driver's license and apologized, stating a neighbor's safe was stolen. Clearly, this was an"Inside job" because nothing else was missing, but the cops had overlooked that fact. No evidence connected me to the crime, so I was released.

Indeed, Upper Arlington youth were granted diplomatic immunity by the local Police. Ironically, on Government Day, my peers appointed yours truly, "Police Chief for a Day." The award was bestowed on the most notorious prankster and delinquent of the 1958 graduating class. Ha.

I mustered all of my prankster abilities from my youth.; I could fix this.

As if my life depended on each word, I said in a controlled tone, "Officers, I call home every day to assure my family that I am okay on the road. This morning, I told them I made it to Jackson, my second call since leaving college in California. My college ID and California driver's license are in my wallet. My home is Upper Arlington, a Columbus Ohio suburb. I was raised by Dr. & Mrs. James O. Watson, Mrs. Watson is my aunt. If I don't report in regularly, she will be alarmed..., and concerned about my safety. Dr. Watson is a prominent surgeon whose patient list includes Ohio Governor James Rhodes among other officials. If I don't ***call home tomorrow***, she will call the Governor. The Governor will call the U.S. Justice Department. And the Justice Department will send Federal Marshalls to my last known location, ***Jackson, Mississippi***. And from what I've observed, ***Federal Marshals*** don't fuck around. So, before you do something against your better judgment, I urge you to consult your ***chief*** on the consequences of the wrong decision. If you think I am making this shit up, tell your chief to call the Watson's residence, Hudson 8-5334, or Doctor's Hospital which my uncle founded." I didn't allude to how my aunt made waves to protect me from arrest when she knew I was on the run from a crazy California man.

"Give me your wallet, kid," the driver said, reaching over to the front seat.

He shone the flashlight on the California Driver's License, then motioned his partner to have a conference out of the car and earshot.

Fearing they would throw the wallet in the swamp after my dead body, I held my breath, when the driver placed the wallet in the pack and walked to my door.

"Get out," he said, opening the door.

Heart pounding, I stepped out, ready to dash into the dark. Suddenly, I am bent over, gasping for breath, reeling from the impact of the GI field pack hurled at my chest.

"Never wanna see your fucking Yankee face around here again."

A moving target for swarming mosquitoes out for blood, listening to sounds of crickets, frogs, and owls in the swamp as dark as a cavern, I followed a dirt road like a blind man in the opposite direction of Jackson. A crack in the clouds periodically reflected moon glare on giant trees draping moss and stagnant ponds. I cussed myself over not familiarizing myself with local roadmap routes before I needed them. I paused and sipped half-filled GI canteen water that tasted like tin. The first sign of hope was the flicker of distant headlight beams darting among the trees. To my relief, an outline of posted road signs appeared at the dirt and paved road crossing. I touched a wooden post and a horizontal metal plate to rule out a mirage. But the car was gone. And it was too dark to read. Fighting sleep, I slapped my face.

A vehicle sped past me as if I were invisible or a dangerous highwayman. I got a glimpse of two posted road signs pointing in opposite directions: **CANTON 20 MILES, JACKSON 10 MILES**.

I walked a neglected 2-lane tar country road toward Canton for what seemed like hours, pausing to sip or splash water on my face. I heard the unique sound of deafening muffled explosions of a speeding car's glass pack dual mufflers and saw headlights racing in my direction. The high beams and loud muffler gave me confidence the driver was young and more likely a good samaritan that would come to the aid of a desperate young hitchhiker. But delusion of benevolence ended when blinding lights and screaming tires instinctively sent me leaping for my life in a dry drainage ditch.

Adrenalin fueled my body rising, one step forward until I caught myself standing asleep at early dawn next to pieces of piano dumped like trash on the side of the road. Its mud-spattered black and white keys stared upward, playing chopsticks to buzzards and passersby. Another face smack, last sip of water, Timex measured the time- time of my impending death: 5:45 a.m.

An hour later, a vintage *Ford* pickup with dead headlight-kids called a *padiddle,* missing a front bumper, and a cracked windshield came into view. The truck bed was loaded with rusted farm implements, milk cans, and bundles of hay. It made a lingering, squealing stop, tingling my spine. The passenger's side door window was halfway rolled down. The driver was an old Negro man with curly hair white as cotton, alert eyes, wrinkled face black as Carney coal.

"Where is ya going?"

"Nearest city.., Canon, I guess."

He waved me in.

I sat in a junkyard seat shedding stuffing, springs poking my butt.

He grasps a shaking floor shift, paused, scanned my face and duds, and shook his head as if having second thoughts. He said, "Son if yuz gonna ride with me, gotta sit on the floorboard. Somebody see a white Yankee boy riding in this truck, they'll lynch us both. Take you to the edge of town. Bus tickets are sold at the country store next to *Sunoco* on Main street."

He read my mind.

I got there safely, thanks to a nameless man with a truck. No thanks to the white police officers in Jackson, Mississippi.

North

First order of business: jettisoning grimy clothes in the restroom. Next, I asked an old white man dressed like a farmer behind the meat counter if this is where I could purchase a bus ticket out of this fucking town. He looked at my dark hair covering half of my ears, clean athletic T-shirts, *Levi's,* and mud-caked *Topsiders.*

He said, "Kid, where ya going."

"North," I said, retrieving $20 from my sock.

Vince picked me up at Columbus, Greyhound bus station. I arranged to stay at his home. Like me, he had graduated in the spring. Only Vince had his career mapped out; upon graduating from Ohio State in pre-law, he planned to join the Peace Corps followed by law school. To his dad's chagrin, his mom Miriam was preoccupied with completing a dissertation in clinical psychology. I found myself sleeping in the same spare third-floor bed I used in my high school and busboy days. Our friend Trent enlisted in the Air Force after flunking out of college. Meanwhile, the Watsons had long moved to a swanky ranch house. I phoned Aunt Helen and informed her of my summer plans and pending induction. I didn't expect or receive an invitation to visit or see my cousins Jane and Jim. I wasn't worthy to associate in the same social circles as his holiness or her elegance.

Vince had a summer job and access to his mom's car. At night we went to movies, and old hangouts like the *Varsity Club* across the street from the Ohio State Football stadium and where photographs of great Buckeye gridiron stars' and Big Ten pennants adorn the walls. During the day I walked two blocks from the Cardis to the Upper Arlington Municipal swimming pool where I biked with Carol Sievers before my California exile. I cannon-balled off the same high diving board that my St. Agatha classmate fell to his death horse playing. Now, eight years later, I followed the same protocol for pool admission of non-members. Pool admission required a paid membership unless one claimed visitor status, wearing a swimsuit under street clothes. I recognized my former high school chemistry, speech, and drama teachers that checked out membership status at the admission gate. I scanned teenage girls in ponytails and pixies sunbathing on beach towels on the pool deck. They lay on their stomachs, bronze legs raised, fluttering for the attention of boys sitting Indian style across the pool, peeking over sunglasses. Others were sitting in mixed groups, sharing magazines, gossiping, and splattering suntan oil on lithe legs and arms.

Time stopped when I saw her climbing into a lifeguard chair overlooking the deep end. Her blonde hair was pulled back, tied in a short ponytail, a white dab of zinc oxide on the tip of the nose. The last time I saw her, she was standing between Vince and an ex-Golden Bear underclassman, Joe Arnold, looking down at the bedridden mugging victim or boxer after the countdown expired: face puffed, black eyes, swollen nose, bloody cotton packs protruding from nostrils. She reminded me of a younger Carol who also had swimmer's shoulders, tan muscular arms, and legs racing across the tennis court, smashing backhand returns beyond my reach. Today, hunched in a highchair, a silver whistle dangling from her thin neck, she surveyed anxious swimmers standing on the pool deck, awaiting the sound of the guard's whistle to dive in.

Surfacing the water below the lifeguard chair, I said, "Remember me?"

Her disciplined eyes affixed on swimmers in the water, she reacted as if I was a distraction until a rest period emptied the pool of swimmers. Then she gave me her full attention. I learned that Joe was no longer in the lineup of her life. Her name was Carol Swanson. She was a blue-green-eyed *Red Cross* Certified Lifeguard and Ohio State sophomore. This was her second full-time summer lifeguard job. In contrast, boys were eligible to be lifeguards at the age of 16; girls had to be 18 to apply. The pool managers were UAHS teachers ranging from middle age to one step removed from a nursing home; one wonders if they could pass the lifeguard test. They were paid more working in a shaded office than the teenage guards sentenced to sit in the hot sun. A former teacher told me that Carol received a *Red Cross* Award on TV last season for saving the life of a young boy drowning in the deep end. She dove in, raised him from the bottom, and resuscitated him. I was soon smitten, I asked her out, apologizing I didn't have wheels.

Not a problem, she said, picking me up in a tar-color pregnant VW beetle-shaped car I'd never seen on the street, in junkyards, or car magazines. Its two-stroke high pitch engine assaulted the ears, blasting death throes under the hood in concert with a smoking muffler that sounded like popping popcorn. According to the manual, it was a *1957 DKW* made in Germany. Carol said jokingly her dad probably bought it at an estate sale in Columbus's German Village because it was cheaper than sending her away to college. A nuisance at the filling station because DKW fuel required mixing a quart of oil in each tank of gas. Mr. Swanson had other priorities than his daughter's higher education: investing in a 140-acre sheep farm in adjacent Delaware County, wherein he spent weekends and annual family's Ft. Lauderdale Christmas vacation. Her mother, Ruth, was a secretary at the *Ohio Department of Natural Resources* and helped pay for textbooks and tuition. Carol worked part-time as a secretary at *Delta Gamma* Sorority headquarters. Carol lived with her younger brother Bill who attended UAHS and was a lifeguard. Her dad was an OSU Agriculture grad, *Dow Chemical* salesman, a proud 32nd degree Mason, and anti-Semite, who despised Catholics and Democrats. They lived in a modest house on Essex Road, one house removed from Star Road adjacent to The Ohio State University Agricultural Department's farm property.

For two months we were inseparable. I double-dated with Cardi to movies like *Westside Story*, picnicked at the river, listened to live music at local nightclubs, played tennis, and watched TV necking in the Swanson's living room after her folks retired to bed. Carol introduced me to the world of synchronized swimming beyond Ester Williams movies. Carol wore a tight swim cap with a chin strap, a nose clip that blocked water and oxygen, and a colorful team suit. She was a member of the prestigious *Carolina Swim Club* that competed against synchronized swimming teams at the Senior National level finals held at the Shamrock Hilton Hotel in Houston. Her team performed in shows across

Ohio. I observed in awe the group's great aquatic skills that demanded precision, and exceptional endurance to hold one's breath underwater executing artistic synchronized routines in concert with the music until their final act: legs disappearing gracefully in a still pool, then suddenly rising, slick bodies shooting surface, heads, and shoulders proudly erect, arms saluting. They mesmerized the audience and the sky.

My August itinerary included visiting my mother at the dreary Ohio State Mental Hospital on West Broad St. I deposited $50 in Kathleen Boyer's account. I met her in the usual visitor area where we embraced like cautious strangers and sat side-by-side on the same wooden bench in front of the laundry building where she sorted clothes. She wore a red bandanna like *Rosie the Riveter,* hastily applied red lipstick and rouge, a denim dress, white cotton socks, and laced leather work shoes. I gave her a pack of *Pall Mall's* which she bolted without pause. I shared my army plans and showed her a wallet-size photo of Carol. In the fog of cigarette smoke and burning eyes, she stared at it and shook her head nervously, rocking, crossing, and re-crossing her pale legs, staring at the sky and her feet. My presence appeared to agitate more than calm her. She clutched the cigarette pack in one hand and held a lit cigarette to her lips in the other. Maybe if we regularly saw her, we could have developed ease between us. It wasn't to be. However, I attributed my irregular visits and ambivalence to the side effects of my passed-around existence. I saw myself as a survivor and beneficiary of my maternal grandparents' unconditional love, and family ties' sharing costs of shelter and education. Our uncomfortable visits would have to do.

In seven years of quiet dining at the Watson's home and reputable establishments, I had rarely conversed with my Uncle J.O. I had not seen him since Jane's wedding and had no inkling why I was breaking bread at a four-star restaurant with him and his lawyer days before my Army induction. He retrieved an envelope from his suit coat, handed it across the table, and explained that it contained a cashier's check for $1800. He explained that this was my share of his son Jim's insurance policy that he had cashed in at his vow of poverty and ordination into the Dominican priesthood. Likely, his sister got the same. After the busboy cleared the table, my uncle retrieved a tobacco pouch and *Dunhill* pipe, scooped a bowl of *Barking Dog* tobacco, tapped it down, and lit it with a pipe lighter that operated like a miniature flame-thrower. I inhaled the sweet aroma that also permeated the Watson library on Onandaga Drive, listening to the advice of my quiet benefactor. "Don't tell your friends about your sudden financial boon because they'll ask for a loan; you'll never be repaid."

Heading west, I exited Upper Arlington again this time by hitch-hiking across the Scioto River at Trabue Rd. The bridge heading west never involved motel stops. Normally, I'd start to arrive in Long Beach within 48 hours via half a dozen cars. The first car that stopped was a dusty black 58' 4-door V-8 *Fury* with New Jersey plates. The white driver rolled down the window.

"Kid, how far you going?

"Long Beach, California."

"Got a driver's license?"

"Yes, sir."

He reminded me of Uncle Mutt but better dressed, wearing drugstore sunglasses and a Yankee ball cap. He did not volunteer his occupation or purpose of his travel other than that the car was an ex-unmarked police car which I believed as it bolted into the fast lane. The interior reeked of rotten eggs and dog food patties served in 2-inch square buns, evidenced in *White Castle* burgers wrappers strewn everywhere. The radio station he listened to was a mix of farm news and country music. He handed me the keys. I was one lucky son-of-a-bitch. Curling in the back seat at a rest stop outside of Indianapolis, he said, "Kid, wake me up at St. Louis." From there, it was all fast food, gas stops, and trading places until Riverside, California.

My last civilian act before reporting to duty was completing a short will instructing my Aunt Marg to deposit $50 each month in my mother's commissary account at the Ohio State Columbus Mental Hospital until my $3300 savings at the Long Beach Federal Savings was depleted.

Fort Ord. Army Training Center

Company B 10th Battle Group 3rd Brigade

It was September 1962. The first soldiers I saw when we pulled up to the training center were square-shouldered male drill sergeants standing tall wearing chocolate brown wool felt campaign hats, starched pressed olive drab fatigues, cuffs slightly puffed, a band tied above black leather spit-polished paratrooper boots with white laces. All wore stern expressions on their clean-shaven white and black faces, their hawk eyes tracking a motley group of America's inductees disembarking a line of *Greyhound* buses' swishing doors: regular army enlisted volunteers, draftees, and six-month reservists which included national guardsmen. The disembarking boys varied in their ages and appearances: peach-fuzz youthful 17-year-old boys to 27- year-old bearded young men sporting flat tops, crew cuts, mop-tops covering their ears, long sideburns, ponytails, rockabilly, ducktails, noodle-like Afros, or parted Eagle Scout, Mormon-missionary hairstyles. Bodies and clothing varied even more. Boys who were overweight, skinny, or muscular dressed in *Levi's*, carpenter's jeans, overalls, *Dickies*. pleated *Haggards*, cut-offs, Bermuda shorts, sleeveless t-shirt, oil-stained work shirt, or silk coconut-button Hawaiian blooming flowers. All kinds of shoes were present: clodhoppers, loafers, ropers, *Tony Lamas*, *Frye* biker boots, custom-made leather sandals, flip-flops, *Keds,* and sneakers which gave clues to class or where the future soldier hailed from. Most carried a small suitcase packed with belongings and rumored fears of drill sergeants' hazing, body surviving eight weeks of brutal physical training, without booze, sex, TV, and the threat of recycling basic training due to injury or illness. The only perspective of real war came from our second-hand experiences with war movies, war novels, and newsreels. The common denominator: *Armed Forces Qualification Test (AFQT)* score verifying inductee is intellectually *trainable, physically fit* e.g., *STD free*, 20/40 vision in one eye, 20/100 in other without corrective vision, and *pain-free* flat feet.

Males born in the WWII era played war as kids in bargain surplus GI equipment e.g., 10-cent shoulder patches, two-bit's cartridge, pistol belts, canteens, $2 field pack, field jacket, two-buckle combat boots, plastic helmet liner, commando knife. A GI's treasure was an $8 leather wool-lined bomber jacket, cap, and goggles. Nazi souvenirs were prized but out of reach for most kids. I was forbidden to take Jim Watson's paternal cousin war hero's classic bolt action German *Mauser Karabiner 98 Kurtz* (K98k) rifle he brought back from the war. Whereas other than a samurai sword and rising sun battle flag, Japanese war mementos were plentiful and cheap. We used *Arisaka* Type 30 bolt action infantry rifles while playing war. But these days of playing war were in my youth. Having not fired a rifle since plinking in the woods in high school, I looked forward to shooting the semi-automatic M1 rifle, smelling the scent of cordite, and lobbing a real hand grenade instead of dust bombs.

The 2nd platoon Company B was composed of 44 mostly unmarried prospective citizen-soldiers representing every race and ethnicity across the land including recent immigrants. Backgrounds were varied: urban working class, farmer, rancher, hillbilly, coal miner, trust baby, goldbricker, prankster, flimflam man, gambler, artist, teacher, banker, gung-ho volunteer, patriotic draftee, and nonchalant *"six-monther"* counting down one hundred eighty days to discharge, and a depressed *recycled* boot-camp loser. Education ranged from a high school dropout to a Ph.D. Our daily schedule was brutal: rise at 4 a.m. speed-shit, shower, shave, dress in fatigues, make our beds- requiring our blankets to be pulled so tightly it would bounce a quarter, jog a ½ mile in fog, complete 25 pull-ups on chin bars. Only then we were allowed to chow down shitty food: eggs over hard, cold burnt toast, slimy bacon, and boiled grainy coffee.

Basic Army physical training (PT) was on par with grueling two-a-day high school football practice in August heat. Together we ran laps, pushed through killer calisthenics and pushups, stumbled in high stepping agility ropes, scaled 15-foot walls, crawled on our bellies in the sand under live machine-gun fire, and woke for night fucking marches in battle gear. Slightly less grueling, we were taught military map reading, measuring the magnetic azimuth using a compass, and how to quickly unload 8-round clips of 30-06 M1 ball ammo clips, our ears ringing shooting *Maggie Drawers* at the rifle range. The scariest experience was standing next to a numskull recruit during hand grenade training who pulled the pin, froze then dropped a fucking live grenade at my feet. My ass was saved by a quick-thinking sergeant who shoved the recruit aside, retrieved, and lobbed the M2 baseball-size pineapple shape iron body like seen dangling on shoulder straps of WWII paratroopers in newsreels, over a concrete retaining wall just seconds before it detonated, a full two ounces of TNT creating a massive explosion. The recruit was discharged on Section 8- Mentally Unfit for service. Addressing the stress of our early morning hustle and barack cleaning detail called for GI ingenuity. My coping strategy included showering and shaving at night before lights out, sleeping on a top blanket rather than under a sheet, and taking advantage of early morning fog to ditch and sneak ahead in the chow line. I admit, I feigned pull-ups, so I could enjoy a leisurely breakfast. By the mid-basic training cycle, the platoon lost three recruits to pneumonia. A dozen, including yours truly, suppressed the symptoms of coughing up the goo by drinking 50% alcohol. We called it *GI Gin* cough syrup.

13 Days

Days drifted into weeks. On the 17th of October in 1962, at approximately 1600 hours, I spotted with some amount of envy a young tall handsome 2nd lieutenant with a shiny gold bar pinned on a utility cap. He was probably straight out of OCS or ROTC. He was observing landmine training that involved probing the sand on our knees with a commando knife for a buried dummy landmine and setting up a *Claymore M18* anti-personnel mine that uses a shaped C-4 charge to fire hundreds of steel balls in a 55-yard killing zone.

Jogging back to the barracks to clean up for evening chow, the avenue appeared unusually quiet. No signs of guide-on leading troops sounded off army cadence or 2nd Platoon Company B 10th Battle Group 3rd Brigade cadre. According to scuttlebutt, our cadre was suddenly transferred and posted at Southeastern military bases with hundreds of GIs around the country. According to the evening news over the radio, a military emergency was triggered by U-2 flight photos of Soviet ballistic missile deployment in Cuba. The 17th of October marked the second dramatic day of the 13-day Cuban confrontation between the United States and the Soviet Union, a face-off for war at sea and sky and a land invasion equivalent to *Tarawa* or *Iwo Jima*.

During this period, recruits were pillared by rumors of sending basic training graduates directly to *Advanced Infantry School,* superseding non-combat *Military Occupational Specialties (MOS*) such as "Military Police" or "Medical Corps." It was commonly accepted that military recruiters' promise of selecting one's own *MOS* assignment was bullshit: KP, cook, grease monkey motor pool, clerk, dog face flamethrower, medic, signal corps randomly awaited us. Six-month national guardsmen are expected to remain stateside and follow their state guard assigned MOS post-active-duty discharge. In the view that our boot camp class represented varied categories of enlistees, at every opportunity, recruits gathered around a barrack radio to hear news bulletins and JFK address the nation. Training courses were suddenly taken much more seriously. We were subjected to cold war propaganda films that graphically displayed the spread of communism on a world globe, with blood-red color graphics flooding maps of Western Europe and Asia. We sought out combat-experienced cadres referred to as *lifers* for updates and advice. They related worrisome stories, stories that had our blood running cold in our veins.

In the first days of the Korean conflict, President Truman ordered reservists to the front line without basic training. Some troops put on planes bound for Soul Korea on Friday were reported killed in action the following week before their next of kin knew they were out of the goddamn country. Why he didn't send those of us that were trained is a mystery to me.

We ventilated our complaints aloud among ourselves: "*FTA* (fuck the army)" during shit details like cleaning commodes, policing cigarette butts, cleaning barracks, and marching in full battle gear. Except for LDS Mormons in their identical padded underwear, recruits did not wear their religion on their sleeve. Politics, ethnic jokes, and racial slurs stayed locked in the footlocker. The most delightful day for Company B 10th Battle Group 3rd Brigade at Fort Ord Army Training Center was not graduation day, but October 28, 1962, sharing newspaper headlines reporting JFK's resolution to the *13-day Cuban Missile Crisis.*

Winter Soldier

No more 48-hour pass, feeling the off-shore breeze, gazing at golden skin *Aphrodite* bodies and long manes of California girls sunbathing in bikinis at Monterey beach. I was now Private 1st class Gerald R. Wheeler who arrived at Tacoma, Washington in a sleeper train car befitting a General. Fort Lewis was home to the 1st Battle Group 22d Infantry Regiment of the 4th Infantry Division. National guardsmen and reservists were relegated to old wooden two-story coal-heated transient barracks built before WWII. Coal dust and soot invaded the lungs and nostrils of all living creatures and mold was embedded in the walls. Reveille was at sunrise in freezing snow. Our army career-*lifer*-platoon sergeant was a sadist born in an igloo. He grinned with delight at shivering wimps pledging allegiance under assault by a great northern and the chill factor grating the recruits' bones.

I buddied with a struggling Hollywood actor who graduated from UCLA and resembled Tab Hunter. He had credits in soap operas and a minor role in *Hawaii Five-O*. We called him Fucko instead of "Dano" the star detective. After morning chow, cleaning barrack and windows, we dressed in long johns, wool shirt, wool overcoat, stocking cap, helmet liner, a steel pot, cartridge belt, and coyote color winter tactical wool insulated waterproof boot, then traipsed in foot-high snow to the outdoor Signal Corp motor pool. We alternated two-hour shifts guarding parked radio vans called *"iceboxes on wheels."* Armed with unloaded M-1 rifles held at sling arms, gloved hands jammed in coat pockets, we would circle a row of vans until relieved for a ten-minute break to thaw out by a wood-burning fire barrel, a fire barrel shared by other soldiers desperately groping for flames in fucking rank order, relegating privates to freezing in the outer ring. I swore I'd apply to OCS, ASAP.

Then we snuck into a van to escape the freezing wind, sitting hunched over a desk facing a radio set and radar screen. Curious, Fucko turned on the power knob. Suddenly iridescent dials and blips appeared on separate screens like the radar screens I saw on the Navy ship scanning for communists invading Long Beach harbor. A speaker crackled and hummed alive. We played with the radio dial searching for an antidote to our present predicament being bombarded with farm news, weather reports, *LAVA* soap commercials, and mysterious two-way radio jabber muffled by static. Finally, an FM station filled the air with a voice and lyrics that abducted my spirit: Roy Orbison's singing, *Crying,* and *Only the Lonely.* Fucko and I sang along in a trance, ignoring green blips flashing on the radar screens, and the sound of rusted hinges of the van's rear doors swinging open. The loud voice of two MPs packing holstered loaded 45 semi-automatics, standing outside looking in, shook us from our reverie.

"Turn that shit off, soldiers, and step outside," ordered a sergeant.

Fucko and I quickly retrieved the M1s, scrambled out of the van, and stood at attention.

"Names?

"Private Wheeler, Sergeant."

"Private Kilmer, Sergeant."

"Well Private Wheeler and Private Kilmer, do you realize you can be court-martialed for abandoning your post?"

"But we haven't abandoned our post," I said defensively. "Our post is the radio van motor pool. We're still here."

"Don't be a smart ass, soldier. Your post is **outside**, guarding **all** the vans, not secluding yourselves in one vehicle to stay warm. Most concerning, and what sent us tracking you down, was this van's radio transmission frequency interfering with the base's airplane control tower creating a potential disaster that could put both of you in Leavenworth."

We nodded and he let us scramble to our posts.

The next morning the Officer of the Day addressed the motor pool guard formation on the protocol of guarding government property, namely prohibiting entering vehicles and "playing" with government equipment.

My face numb, lungs exhaling vapors thick as fog, I was going through the motions when I heard the sadist platoon sergeant's gruff voice call out my name during the reveal, "Private Wheeler, report to Company HQ immediately after morning chow."

"Yes sergeant," I snapped back, wondering if this was added punishment for the van shit. Yesterday, the motherfucker ordered us to give fifty pushups in the snow. Fuck me.

The company headquarters was the leadership and command bureaucracy keeping direct tabs on 175-weekend warriors on temporary active duty in regular Army purgatory. The Company Clerk was the first person any visitor encountered inside a small building. He sat behind an olive drab metal desk, fumbling with forms stacked in In and Out baskets. A glowing wood-burning stove stood on the right side of the room, between the office doors of the Company Commander (CO) and Executive Officer (XO) respectively. The Commanding officer, a Regular Army Infantry Captain, and XO, 1st Lieutenant must've drawn the short stick to be put in charge of a unit composed of six-month jerkoffs and admitted legal draft dodgers with only a few months left of active service. The walls were decorated with photographs of the Chain of Command: Base Commanding General, Battle Group, and Division Commanders, and a table of organization (TO) wall chart of our company. The unit flag and old glory served as bookends to the desk assigned to the First Sergeant who set the tone of the unit. As base outliers, we got little notice from GIs intransient to and from Germany, South Korea, Japan, and other foreign Army posts. A few enlisted lifers and career officers at Ft. Lewis expressed that working with six-month enlistees was easy duty because many were college-educated, more mature, and tolerant of army discipline culture imposed in months not years. The only disciplinary incident I witnessed was the day a PFC reservist that was a trust baby with a BMW personal car, refused to stand in line for E-1 $82 monthly pay he referred to as "chump change." He quickly changed his mind after the payroll officer threatened him with a court-martial.

The Company Clerk was blond with light skin and vivid blue eyes, He wore starched fatigues and spit-polished boots. The burning wood crackled and sputtered like a campfire. Here in Washington, there was no scent of coal or dust-rotted curtains. The snow melted on my boots, creating puddles, I removed my helmet liner, and gloves and unbuttoned my overcoat. Anticipating a shit detail worse than frostbite guarding fucking trucks I said, "What's up, Corporal."

He retrieved an *Army Personnel 201* file with my name on it.

He said, "According to your file, you can type."

"Yes, Corporal. Taught on blind keys by Holy Cross Brothers." This trusty line worked before. Might as well try it again.

"We got several college guys with typing skills. But you have the first crack."

"Why me?"

"We got one thing in common."

"What's that?"

"I'm a *Forty-Niner*. Class of 60"

"No shit."

"No shit. Nineteen January discharge. Captain wants a competent replacement. Great benefits: first at morning chow, no reveille, GI-ing barracks, guard duty or work details."

I was granted a 14-day pass before reporting to my new assignment. With no family to visit, I touched down on Christmas Day 1962 at Fort Lauderdale Airport after landing a direct flight from Tacoma to visit Carol Swanson whose family joined the snowbirds renting condos near Florida's white beaches. Seeing a pale face GI dressed in Class A uniform sitting alone in the back of the plane, a pair of beautiful TWA flight attendants took pity, escorting me by the arms to the first-class section, plying me with Champagne as their patriotic duty. It was the beginning of a long romantic rendezvous: sunbathing on sugar-white sand, reading paperback novels, necking in an empty rented canopy, and moonlight strolls gazing at a regatta of sailboats draped in Christmas lights.

Two weeks later, I awakened, staring down at the snow-capped summit of Mt. Rainier. During the remaining Army days, I escaped the winter elements, feeding cordwood in HQ's stove, and completing daily Morning Reports listing dwindling names of the Company's roster. By mid-February, I completed final rites of discharge passage: collecting $45 mustering out pay, after two-day KP duty peeling a mountain of Idaho potatoes, crippling my fucking fingers.

The Road To Public Service

Inspired by JFK's call to public service, I contemplated applying to the *Peace Corp* but opted for transferring from the California National Guard to its Ohio counterpart. Flush with three thousand dollars, I planned to buy a used car in Long Beach, drive to Columbus, court Carol Swanson and re-establish roots in Ohio.

Estranged from their runaway married daughter, Marjean who was living in Oregon, and their divorced Air Force vet son Bill, whose last address was a desert trailer in Arizona, the Paskells had moved from their former home in Huntington Beach. I caught up with them before my cross-country expedition.

On Valentine's Day 1963, I shelled out $850 for a cherry red 1954 *MG TF* roadster. It reminded me of high school senior classmate Arty Wolfe's red MG-TD roadster he buzzed around Arlington in, picking up girls. I pictured Carol Swanson's raised eyebrows and excitement when I pulled into her Essex Rd driveway, top-down, blowing the horn. The *MG TF* model has dual carburetors, spoked wheels, a spare tire on the rear luggage rack, a soft top, side curtains, embedded headlights, fog lights on the front chrome bumper, leather bucket seats, and a wooden panel. Under the driver's side of the hood was a nifty tool kit secured in a metal box. I packed the car to the brim, intending to leave alone for Ohio when Ernie asked for a ride to Columbus to meet his sister and visit their brother dying of cancer. I agreed.

Grimacing from butt pain from the thousand-mile hard drive, (I later learned that my MG's suspension system was missing a rear leaf spring and had a rusted front coil spring set), we pushed through the torturously long drive when suddenly steam and water started spewing outside the radiator like *old faithful*. I pulled over and shut off the engine. According to the map, we were miles from Oklahoma City but not from hope. We had the foresight to carry a desert water bag on the luggage rack. A tool kit was under the hood. And Ernie, a Navy-trained submarine diesel mechanic, called upon to fix leaks during depth charge attacks, quickly identified the problem, removing a cracked radiator hose. He surveyed the landscape of cornfields and livestock for barns, silos, and tractors because farmers are used to jerry-rigging farm machinery equipment.

An hour later we refueled at a *Texaco* and replenished the desert water bag. We never discussed our history or his parenting skills. Political discourse-wise, I related my positive impression of JFK and regret over being too young to vote for him. I mentioned my surprise at how physically short Nixon was when he made a presidential campaign stop at Long Beach State in 1960. Ernie volunteered he did not vote for Democrats, and especially disliked Roosevelt because he ordered farmers to slaughter and bury their livestock to drive up the meat prices during the depression. I offered no rebuttal like alluding to Keynesian economic theory, the ramifications of FDR's New Deal policies, or the Truman Doctrine and Marshall Plan, issues discussed in college classes but rarely continued in the quad because students were mostly focusing on their next test, surfing, sailing, skiing, and getting laid at the next wild beach party. In the years I lived with the Paskells, I never saw Ernie reading a book or newspaper article other than the sports page. I was surprised to find out he knew of world events.

It was too cold to make a top-down entrance. I exited a mud and sleet-splattered *MG* as Carol descended the front porch. She appeared more stunning than our last embrace at the Columbus Airport the previous month. She usually shunned wearing hats. But today she wore a beige faux fur Russian Cossack style hat that smothered her blonde hair, but not her face and eyes that were touched by nature's colors and youth's optimism. A plaid scarf casually wrapped around her neck matched a short, pleated skirt exposing pale knees and sexy thighs above high brown leather boots with western heels. The Swanson's modest home was not conducive to young lovers: no basement recreation room or guest bedroom. The downstairs bathroom was the size of a TWA urinal. The only opportunity of being alone was in the living room when everyone was asleep. Her mom's work schedule, 7 a.m.-2 p.m. as executive secretary of the State's Natural Resources Director, ruled out the afternoons, her brother, Bill was in high school and often stayed up late. We drove to a coffee shop at Lane Shopping Center. I told her my plans to share a basement apartment with Cardi in the converted house off the OSU campus. Cardi was starting law school. The priority was exploring County and state job opportunities working with youth and transferring to the Ohio National Guard.

Cardi lived in the basement of a three-story depression-era mansion divided into four one-bedroom student apartments. The top floor had an iron sun iron deck and fire escape at the rear with steps leading to the backyard. Three upper-floor apartment tenants shared a communal bathroom on the second floor. The singular advantage of basement tenants was a separate bathroom and shower. The shower was a crude nozzle suspended on a garden hose over an open drain separated from the kitchen area and commode by shower curtains. The only natural light in Cardi's abode was two basement window wells filled with dead leaves and debris. Living in the basement was like living in the bowels of a vintage gunboat. The ceiling exposed water pipes and heating ducts with sounds of their own. The bedroom contained a double bed and rollout sofa, which left no space for standing when both were in use. The kitchen had a two-burner stove, two flea market chairs, and a peeling Formica table that stood with a limp. The remedy for the dust and dank smell was a wobbly ceiling fan missing a blade and open windows designed to stay shut. But my $25 monthly rent share was a bargain to a homeless man.

In high school, on government day, I observed a juvenile probation officer, so I explored the state and county youth service positions once I arrived in Ohio. But I hit a dead end. I was one of the hundreds of liberal art graduates with no traditional job experience. I interviewed for a teaching job at a parochial country school and was interrogated by a priest on my attitude toward abortion and responded to a civil service test posting for *Franklin County Welfare* Department Caseworker which took a month to process and applied to *Ohio Youth Commission* that had no current openings or parole officers. Meanwhile, the 1954 *MG TF* roadster repairs, car insurance, and $10 monthly alley garage rent were draining my dwindling savings. Fortunately, a carryout grocery and liquor store at the corner of High Street across from our basement hovel needed a part-time cashier.

Vince's law school classmate, Bradly McGibbon was a fast-talking second-year law student with a photographic memory who never went to class. His face was a photo image of Teddy Roosevelt's Rough Rider; wrinkles working outdoors tending family farm, deep musings dissecting inane subjects to hear his voice. He had the body of a linebacker ready to pounce on any side of an argument like his county judge father, and the eating manners of a caveman. But his Spartan apartment's loss of a bathroom was made up in space for two double beds, a standing room, a TV, a reading chair, and telephone, and a kitchen conducive to guests. The first time I met him he was wearing a U.S. Army officer's class A uniform with a shiny gold bar on the epaulets designating the status of 2nd lieutenant in the Ohio National Guard.

The timing was fortuitous. I had a month to activate a transfer from California to the Ohio National Guard. I contacted the State Guard headquarters and was ordered to report to the 112 Medical Corps Battalion Headquarters at Fort Hayes in Columbus. The unit had a slot in the upcoming Officer Candidate (OCS) School that involved a year of intense monthly weekend classes instead of standard drills beginning in October and two sessions of two-week intense OCS summer field training with pay grade raise from private E-1 class to sergeant E-5.

Carol was a full-time Home Education student, living at home, a member of Delta Gamma Sorority and part-time paid secretary at their National Headquarters across from the Scioto River Rd, bordering Upper Arlington. Notwithstanding her challenging study and working schedule, we managed a few dates going to movies, and parties and being alone at night at Bradly's third-floor apartment because he was sleeping downstairs with a psychology graduate student. The basement bungalow was too dank for us lovers. But it did not deter Cardi from maintaining his lady's man reputation.

The Welfare Trap

In September I moved upstairs for the same rent plus a share of the phone bill and commenced my career in public service as a $320 a month plus mileage Caseworker at The *Franklin County Welfare Department* making home visits to General Relief Public Assistance recipients. The department's sprawling office was crowded with small metal desks occupied by mostly white female clerks and college-age caseworkers interviewing welfare applicants, and staff recording home visit contacts in *Dictaphone* in sound-proof booths. I was one of the new young beneficiaries of the *1962 Public Welfare Amendments to the Social Security Act* engineered by JFK that allocated funds for local public welfare agencies to provide casework and job training to welfare recipients. President Kennedy was moved by Michael Harrington's book, *The Other America* which raised public awareness of poverty hidden under the concrete maze of sky-high freeways overlapping ghettos and people displaced by urban renewal and racist housing policies.

Sitting at a desk next to me was a young attractive college-age Black female caseworker named Jeanette Carrington. She introduced herself as the person responsible for orienting new hires which involved ride-along home visits, checking documentation, clarifying Public Assistance policy guidelines, and community social services agencies. Having recently completed 1 year of service, Jeanette was classified caseworker II with a step raise and transferred from a General Relief caseload to Aid to Families with Dependent Children (AFDC). She was from Athens, an Art history Gene at Ohio University, class of `61.` I told her I graduated from Long Beach State, was born and raised in Columbus, of course skipping the details of who raised me, cognizant that the chances of an Upper Arlington resident applying for financial assistance in this county were the same as him being Ohio State Penitentiary convict applying for parole.

She described the plight of high school dropout single mothers losing assistance tied to a dependent child if their biological father were caught at home or not reporting income from absentee father or her part-time work. According to Jeanette, unskilled single Black mothers were trapped in the culture of poverty by welfare policies and racial discrimination. While AFDC was a lifeline for dependent children and financially disadvantage single mothers, welfare policies created traps to escaping poverty: a disincentive for absentee fathers to contribute income or visit offspring, encouraging mothers to lie about sexual partners, depriving children of paternal discipline, and offering minimal coordinated social services to improve single mother access to education, and training. In the final analysis, Jeanette asserted AFDC families are doomed to live on subsidence level beyond the imagination of middle-class Americans and the well-heeled like the Watson's where Aunt Helen hoarded boxes of *Hormel* chili, *Campbell* soup, fruit, toilet paper, toothpaste, and soap in padlocked walk-in fruit cellar since WWII, and stored prime cuts of beef in a local private meat locker. And Grandma Carney having raised a brood of nine children on a farm in the worst of times was a culinary artist. She ran a tab at Eastend's mom-and-pop grocery store and paid it off monthly with her combined Social Security check for widows and Uncle Walter's never mentioned monetary support.

General Relief recipients were generally the elderly with health problems or married to disabled or laid-off blue-collar spouses that depleted savings and were ineligible for Social Security. They relied on Government Cheese, church handouts, and food banks. They lived in drab duplexes like Grandma Carney's in Columbus Eastside before she died. A few tended a vegetable garden next to a trailer in a rural township that had more tombstones than houses. I violated department policy against transporting invalid recipients to the hospital to visit their kin and picked up blocks of neon yellow color Government cheese. Many GR recipients expressed humiliation over losing their self-reliance and dignity like their parents they witnessed as children of the *Great Depression*. Occasionally, I would knock on the door of a low rental apartment and a depressed childless couple answered the door in a drunken stupor in a smoke-filled room. After making unannounced AFDC home visits in ride-along with Jeanette, I concurred with her impression that some white families met the pejorative description of "white trash": slovenly appearance, floors strewn with trash, empty beer cans, cigarette butts overflowing ashtrays, dirty dishes, children's toys, and screaming toddlers ignored. Whereas young overwhelmed Black mothers dotted over infants and made sure school-age boys' clothes wore clean and girls' hair meticulously braided for school. Unwed mothers with multiple children by different men that absconded or were incarcerated, struggled *to* make do alone on meager government aid. The primary family backup guardian of mothers incapacitated by drug or alcohol addiction, or mental illness were grandmothers who were even less able to support the children of their children.

Until I met jobless and disabled adults facing hunger, health problems, eviction, and surviving on *Government Cheese, Food Stamps,* and charity, my perception of poverty was abstract or limited to images portrayed in movies like *Grapes of Wrath, Tobacco Road,* and books like, *Let Us Now Praise Famous Men*. The only mention of hunger in childhood was when Grandma Carney eyed a morsel on my plate, she'd say, "Jerry, people are starving in China," which I never took seriously. Regardless of guardians in passed-around existence, there was never a notion of going hungry, wanting necessities, toys, or entertainment. I have vivid memories of adventures at Carney homestead, and Logansport good days learning from my stepfather how to ride a two-wheeler, bat left-handed, and explore caboose and steam engines. However, unlike the psychologically repressed trauma over sudden separation from maternal grandparents and migration from New Lexington to Logansport, I could not escape reacting to my mother's psychotic episodes during stepfather Earl Boyer's absence. I was ill-equipped to protect her from self-destructive behavior like drowning in the bathtub, running around the house half-naked, or escaping into catatonic withdrawal. The nuns and Earl's folks intervened.

However, even they could not solve the problems brought on by mental illness. My mother disappeared and returned and disappeared again and again from our lives. This created a period of profound anxiety and unpredictability like riding an emotional rollercoaster for me. A ride that ended the dramatic September day a priest suddenly whisked me out of second-grade class, drove to

Logansport train station, and put me on a Columbus-bound train car packed with Army heroes and brought me stability. Many of the families I visited were not so lucky. No priest intervened on their behalf. No wealthy family members took them in. It was up to me to support these children.

The closest I got to *The Other America* as a kid was living in Eastend Columbus in summers in the early 50s while Aunt Helen's children went off to private camps. Unlike the all-white population in Upper Arlington and Grandview Heights, Eastend Columbus was a diverse working-class community and a socioeconomic notch below Grandview Heights. Grandma Carney's relocation was necessitated by Uncle Joe's sudden death, leaving his wife a widow with a severely handicapped child. I lost contact with them as a teenager and an adult, but later I learned they were supported by Uncle Joe's older brother, Walter Carney. On Saturday afternoons, in Grandview Heights, seven-year-old boys walked alone a mile uphill to *Grandview Theater*, paid two-bits for the double feature, cartoon, bag of fresh popcorn, and chocolate *Holloway* sucker that took two hours to consume, and hiked on Sunday in *Grandview Woods.* Preteen Eastend boys fended for themselves by trash hunting in alleys, crafting toys, playing endless games at *Roosevelt Recreation Park*, sneaking into movies, and ogling dirty magazines. They also attended free *Columbus Redbird's* double-headers as members of the *Knothole Gang.* The word "poverty" was never part of the lexicon. Most of our entertainment was inexpensive if not free. Kids walked, rode makeshift bikes double, cashed in nickel deposit bottles, bartered, borrowed, crafted toys, and spent spare pocket money on jawbreakers, licorice, and suckers. Our owning a family car was an exception; most Eastend kids did not come from families who could afford cars.

November 22, 1963

"The President's been shot," Jeanette cried, staggering out of a supervisor's office in tears.

I was dreading dictating the week's home visits. Suddenly the teeming office was struck by a silent pall, flashing phones unanswered. Staff froze in their chairs, called loved ones, and gathered around colleagues listening to minute-to-minute radio news reports on the status of the mortally wounded President shot in the presidential motorcade through Dealey Plaza, Dallas, Texas.

Hope died Friday at noon in the authoritative voice of Walter Cronkite, "President Kennedy died at 1 p.m. Eastern Standard Time, in Dallas Parkland Hospital." Franklin County Welfare Department immediately released employees into an atmosphere of a funeral wake. Staff walked out of the office like zombies. I drove directly home and knocked on Cardi's door. No response. He probably was at the law library. Carol was somewhere on campus. I went upstairs and called my roommate at his girlfriend's apartment below where he spent most Friday afternoons. He answered hurriedly as a man interrupted typically did. I said, "Bradley, President Kennedy was assassinated."

"Later man," he snapped. "I'm busy."

I heard a female voice moaning in the background. I slammed the phone down in disbelief, reflecting that today was worse for me than the Cuban Crisis and the police car ride in the Mississippi swamps. My fucking roommate can't take a respite from getting his damn rocks off.

In 1964, I was introduced to the Social Work profession in the TV drama series *East Side/West Side* starring George C, Scott. His role advocating social justice for Harington's *Other America* went beyond social workers verifying the poor's living conditions via home visits to justify government *Food Stamps*, crowded *public housing,* free *Government Cheese,* and financial pittance; a subsistence existence that kept families in place at the lowest rung of the socio-economic ladder. Scott portrayed grassroots social workers engaged in mentoring the poor in strategies of community organization and social action to hold bureaucracies' established leaders accountable for ignoring their plight.

On Groundhog Day. I responded to *Ohio Youth Commission's* call for Juvenile Parole Officers (PO) for adjudicated delinquent youth committed by courts to juvenile correction facilities ranging from 100-bed youth camps to the nation's oldest (1857) and largest (1500) residents open setting reform school, *Boys Industrial School,* Lancaster, Ohio—the birthplace of William Tecumseh Sherman. I was one of six POs assigned a rural caseload of 75-100 parolees. Our group of POs was composed of white liberal arts Genes: ex-teachers, veterans, and conscientious objectors that saved lives in third-world countries devastated by natural disasters. We reported to Eugene Penney M.S.W. OYC's Columbus Field Director. The monthly salary was $380. The benefits included the oldest *Chevy Impala* in the State's motor pool, pair of standard handcuffs, and a blank arrest warrant to take into custody a juvenile parole violator.

Monday through Thursday were "field days." We tracked down absconders in Central Ohio and made court appearances pleading second chances for incorrigibles, shoplifters, petty thieves, glue-sniffers, drug and alcohol abusers, and joyriders. Youth adjudicated delinquent for violent crimes such as armed robbery, assault and murder were committed to *Training Institute of Central Ohio* (*TICO*) a closed modern correction facility located in Columbus and staffed by youth leaders, clinical social workers, teachers, a psychologist, and consulting psychiatrist. Female delinquents were placed in *Girls' Industrial School* a.k.a *Scioto Village for Girls)* located on the west bank of the Scioto River, Delaware,

PO field days began in the early morning, driving miles across county lines to counsel parolees at home, school, and worksite, talking with school counselors, family clergy, and military recruiters. We rarely used handcuffs, unless a small county sheriff's mandated cuffs as a condition of jail release. Our second chance recommendations were often rejected by elected local juvenile court judges. Reality dictated that ninety percent of violators POs formally recommend placement in a *residential treatment* setting meant a return to the infamous *Boys Industrial School* described as a *warehouse with a revolving door.* Our written recommendations reflected the boss's social work professional ethics: society's *throwaway children* whose dysfunctional behavior was attributable to parental abuse and neglect, inadequate schools, and poverty deserved the same residential treatment options available to troubled children of affluent families. Thus, we spent office days writing fictional rehabilitation recommendations for non-existing public residential placements described in glitzy brochures of expensive private psychiatric settings like the *Devereux School* in Pennsylvania. The judges ignored our recommendations. Like many young middle-class college-educated welfare caseworkers that never knew hunger or faced eviction, and supported welfare rights, young POs had a strong commitment to keeping our boys out of jail for petty crimes. Many POs as teens didn't get caught or were counseled and released by police. We knew what a difference this made in our lives turning out how they did.

There were days of irony and comedy in transporting violators to facilities. An example of the former occurred while driving a 15-year-old, long hair "joyrider" from a Nelsonville jail to BIS. I had to get his help in resuscitating the disabled State car, bringing the smoking engine in its death throes, ten miles from BIS back to life. Finally, after two tows to the motor pool, I convinced OYC bean counters that POs could opt for transporting wards in personal vehicles and paid mileage. Given the *MG TF* roadster's chronic maintenance problem, single passenger seat, and leaking curtains, I sold it for the same price ($850) I bought it and made a down payment on a 1964 oyster white manual shift VW Beetle with an accordion sliding canvas (rag-top) sunroof, shoulder strap seat belts, and a reserve gas tank.

The most unconventional field assignment involved retrieving female *Scioto Village* AWOLs detained at Williams County Jail, Bryan, Ohio, 160 miles north of Columbus. It was against department policy for male POs to transport female parolees, without a female co-worker. Unfortunately, female POs were attending mandatory training in Cleveland. After a three-hour drive, a pudgy tobacco-chewing Deputy County Sheriff introduced me to the trio of 15-year-old runaway female wards of the court he called "the wild bunch," cuffed shoulder-to-shoulder to a church bench outside his office. The first girl was a copper tone tanned skin peroxide pixie named Becky a.k.a. *Blondie*), one eye brown, the other green, and a pug nose. She was wearing a psychedelic T-shirt, nipples pressing a life of their own, a mini denim skirt, and leather sandals I had seen on hippie girls in San Francisco. She was linked to a heart-tattooed arm attached to a skinny girl named Henrietta a.k.a. *Amber* with an orange ponytail, amber eyes, smooth skin, dimpled chin, parakeet beak, budding chest, dressed in sheer pink blouse with missing buttons, bead necklace trapped in her cleavage, tight jeans cutoffs, and flip flops decorated with a plastic rose. The third case file read Nancy a.k.a. *Peaches*). She had stringy yellow hair, gold-brown eyes, rosy cheeks, a perfect nose, and a swimmer's shoulders. I displayed warrants, signed release papers, and for extra cuffs, the Deputy Sheriff insisted the runaways wear as flight risks.

Amber and Peaches linked together, squeezing into VW's rear seat. Blondie jumped in front, holding a bag of personals in cuffed hands. I started the engine, Peaches pulling Amber's arms between seats, turning the radio dial from Johnny Cash to White Rabbit.

Retrieving a pack of *Newport Menthol* cigarettes from a bag of personal items, Blondie asked, "Gotta match, good looking?"

"Negative." I said, reaching for the gear knob, grasping Peaches' free hand shifting from neutral to my crotch, protesting, "Missy, that ain't first gear."

Shifting gears, I feel fingertips brushing my burr hair, something slimy licking my ear. I looked at images of gleaming amber eyes, thin lips, and wiggling tongue in the rearview mirror.

"That is," I said in a raised voice, braking on the shoulder of the road, threatening to return to jail. During two pit stops, I plied them with burgers, shakes, and fries, listening to their life stories, fantasies, and lies. Between the lies that downplayed the abuses of the adult world, I heard the truths that drove them to the mean streets. They were surviving the best they could.

In the following months, I completed intense OCS 48-hour weekend courses in Map Reading, Logistics, Infantry Tactics, Leadership & Command, taught by Ohio Army National Guard senior officers. Drill & Ceremony in the field involved *Tactical* training *Officers* bellowing marching formations on the run at platoon size class, conspicuously tapping the helmet liner of the cadet out of step with a swagger stick; humiliation echoing in the outlier's brain. This was followed by randomly selected cadets exchanging places with *tact officers,* and leading marchers, often causing pandemonium and ridicule. I broke out in hives. After surviving two weeks of grueling physical OCS summer camp training in 63` and 64`at Grayling Michigan, I was commissioned 2nd lieutenant in the Ohio National Guard.

Fellow PO Ron Studebaker was a year older (24) married and a conscientious objector. I erroneously assumed that conscientious objectors were anti-military, unpatriotic draft dodgers afraid of risking life in combat or hated guns like Private Slovick the only GI executed as a deserter in WWII. I was admittedly a WWII/Korean war-era kid that fantasized about my absconding father dying a war hero. I was addicted to war movies, playing war, a proud Carney descendant of a Civil War Union soldier that fought in the Battle of Gettysburg. I learned to shoot guns before I could read, enjoyed plinking, hunting rabbits and varmints; only sorry for murdering birds to prove my marksmanship, a junior gunsmith by age 14.

My attitude only changed the day I accidentally shot my cousin Bill in the hand with a pellet gun loaded with a BB as he was pointing at a crow flying overhead. And years later, I confirmed the attitude shift at Ft. Ord basic training, lugging 9-pound M-1 carbine and shooting *Maggie's Drawers,* plagued by ringing in the ears.

Ron was one of five brothers and three sisters raised on a 140-acre milk cow and hog farm in Englewood, Ohio. He was the first of three brothers that served as conscientious objector volunteers in third-world countries: Morocco, Cambodia, and South Vietnam. He attributed the decision to change his Selective Service status to a myriad of factors: Christian values of humility, charity, and empathy for the sick and poor instilled by his parents. He said: "My mom taught me to listen to your heart, which means listening to your emotions to what is right. My dad farmed most of his life and taught us practical things-how to do things and perseverance will bring rewards. Family ties with the *Church of the Brethren Volunteer Service Program* and faculty at Manchester College, Indiana developed a strong interest in the concepts of open dialogue, hands-on service, and reconciliation as a means of breaking down differences of opinion and outright resolution of conflicts other than violence. These differing attitudes and opinions helped me clarify the decision to serve in an approved alternative program instead of carrying arms in the military. The review process took two grueling years. It kicked off a period of losing close friends that called me, "Chicken and a Yellow Belly Coward."

In 1960, Ron took a break from college, completing eight weeks of training and orientation through the *Brethren Commission* at Windsor, Maryland, six days of rocky sea travel, and began assimilation to the culture and language with family in Kassel, Germany, in preparation for the assignment to an orphanage project. But he learned from *BBC Radio* that a Gene earthquake occurred in Morocco, North Africa. He wrote in his journal:

I immediately called the Brethren European Director and requested a new assignment. Within 24 hours I was in Agadir, Morocco scared to pieces but ready to do whatever I could in that chaotic situation. It was a 12-second earthquake occurring at 11:41 p.m., registering 5.7 magnitudes on the Richter scale. It obliterated the city of Agadir, killing more than 12,000 people, and leaving 35,000 homeless. Death, destruction, fire, panic, injury, mass confusion, uncontrolled emotion, searching for family and loved ones....all these elements were pervasive and in-your-face. In addition, this quake happened during Ramadan, which injected more issues because eating during daylight is prohibited. Such needs included:

. Extracting people from the rubble

. Providing food and/or water to those still trapped

. Consoling children and removing them to temporary tent shelters

- Stacking the dead to be trucked to bull-dozed trench burial sites
- Unloading, unpacking, and distributing medical supplies
- Help post names and status on information boards for next of kin to review

Every night for two weeks, I worked with the French military, tapping on pipes, poles, and concrete structures listening for noise or feedback. Any positive responses were marked with white spray paint for rescue work at daybreak. Larger NGOs concentrated search and rescue operations in populated urban areas, leaving the smaller service volunteers like the *Church of the Brethren* to organize small international teams to search for victims trapped in isolated mountainous areas only accessed by foot. I cajoled the US Naval Base staff in Kenitra, Morocco to donate building supplies to construct a medical clinic and a pickup truck to deliver food, water, and medicine. I delivered a baby girl in my truck's front seat and wrapped her in a bath towel. The most intense experience and challenge were amputating a gangrenous leg above the knee of a woman trapped under a boulder with a hacksaw sterilized with alcohol, without anesthetic, medical tools, and only elementary first aid training and a rope tourniquet. Another rescue I vividly remember was stitching a lactating nipple of a mother's injured breast with a needle and thread while a British volunteer held her infant.

Comparing Ron's history of working his way through college and dangerous humanitarian service to my risk-free beach college life drinking cheap booze, easy jobs, and family largess, I formed profound respect toward a new friend. We exchanged work stories fighting for kids raising themselves without adult role models, lack of community resources like *Roosevelt Recreation Park,* school vocational, and remedial education programs.

War On Poverty

In 1964, President Lyndon B. Johnson implemented *The War on Poverty.* Its goal was to lessen the number of families living in poverty. One means of achieving the objective was establishing programs that train the disadvantaged and at-risk youth with integrated academic, and social skills and vocational guidance and teaching so they can get quality and long-term jobs. Federal funds were directed to state and local public and non-profit social services agencies to mobilize and better coordinate community resources for impoverished families and troubled youth. Such a mission required expanding trained staff. In this regard, the United States Health, Education and Welfare Department (H.E.W.) offered no-strings grants to attend university-accredited two-year master's in social work (M.S.W.) programs. Ron and I recognized we were low-level welfare caseworkers with cuffs.

At 1200 hrs. on Saturday, July 11, 1964. Colonel Good, Ohio National Guard 137th Medical Battalion Commanding officer released 2nd Lieutenant Gerald R. Wheeler, Ambulance Platoon Leader to leave the weekend drill. I had one hour to race home, change from class A Army uniform to a rented tux, drive to *First Community Church* in Upper Arlington, recite the marriage vows, kiss the bride, then commence a three-day honeymoon at Cedar Point, Sandusky, Ohio. Carol's younger brother, Bill, was the best man.

The newlyweds moved into a $50 a month, coal-fueled, gurgling steam-heated, second-story one-bedroom apartment above a children's dance studio on Arlington Ave across the street from Upper Arlington City Police & Fire Departments. Carol was a senior at Ohio State, working summers at Delta Gamma Headquarters. The apartment was furnished with second-hand furniture and a refrigerator with a broken freezer. Carol sold the DKW, using the proceeds to purchase a stereo, 12-inch TV, and a pure-bred female German Shepherd puppy we named Braun who had the temperament of a golden retriever and paws of St. Bernard.

Mentors

I received a no-strings H.E.W. $2000 annual stipend plus books and tuition at Ohio State School of Social Work. Ron accepted the University of Connecticut's financial aid package conditioned on two-year employment in the State's Youth Services.

The 1966 Ohio State M.S.W. class was predominantly white females. The only Black student, Alan Holloway, was a superintendent on leave from OYC's Boys Camp in northern Ohio. In his mid-thirties, married, and father of one child, he was looking to relocate his family. We shared interests in football, sports cars, and juvenile justice reform. Al volunteered that he was an Army veteran that survived nearly three years of captivity as a prisoner-of-war (P.O.W.) in the Korean War. I never probed into the suffering details reported in the news: starvation, freezing, torturous interrogation, and isolation that drove GIs to wither away. Al and I quickly formed a bond.

Having laid tons of concrete driveways and basements in duplexes adjacent to my high school and expanding the Upper Arlington community, I offered Al assistance in apartment hunting. But no landlord in the 43221-zip code accepted Al's check or cash. His patriotism, military sacrifice, and professional qualifications were dismissed by the color of his skin. This raised the specter of my passivity in fighting racism after the police encounter in Jackson Mississippi in the summer of '62. The ingratitude I exhibited as a child toward Lizzy the Watson's maid, who kept my hidden menagerie secret, cleaned my messy third-floor sanctuary and bathroom, changed sheets, and washed dirty laundry that I conveniently dropped in a chute landing three floors below. These clothes magically reappeared after being meticulously ironed and folded, lugged three flights of stairs to dresser drawers by her. My past failure to befriend Black dishwashers and busboy peers who were denied higher education options and exploited by white restaurateurs. My shame over **not** joining the civil rights movement after witnessing the courage of an elderly Black junk collector that risked lynching for abetting a young naïve Yankee's escape from Mississippi's violence and not investigating why Ohio National Guard brass excluded Blacks in its OCS class, came to the forefront. I was still seething over Upper Arlington City's racial discrimination policy forcing Al' to resettle his family in a rundown duplex near the OSU Law School that ironically professes *Equal Justice Under Law*. But I didn't know what to do to fight the institution at large.

In the fall of '64, The Ohio State School of Social Work's faculty was as white as all my teachers since St. Brigid's at Logansport through college had been. Unbelievable! The graduate student Orientation was a bus tour led by an elderly tenured, twice married female Freudian clinical social worker whose first husband, according to gossip, shot himself in their bedroom, and her second mate disappeared. Standing next to the driver, facing wide-eyed students, she directed our attention out the windows at Columbus Eastside's landscape "sliding into plight" where I played a summer before my California exile and Grandma Carney's wake. As she spoke, gazing at passersby, one hand stroked the air like a priest giving her flock a final blessing of poor souls; her words fading as old notes soon lost in her dementia. Next, a drive-by of massive century-old stone walls of the *Ohio State Penitentiary* where I attended mass as a child with a high-stepping maternal aunt served breakfast by god's notorious sinners praying for parole. And lastly, a gander at *Ohio State Mental Hospital* on Columbus Westside, the residence of my biological mother, Kathleen, and her younger sibling, my Aunt Wanda since 1946 with its famed dungeon tower accessible only by a set of spiral stairs I climbed years before. These stairs led to lonely dreary wards that I dreaded to ascend because they triggered flashbacks of childhood trauma of sudden separation from maternal grandparents, images of my mother's sorrowful face of psychotic episodes in Logansport days, and my rage reflecting on name-Robert Vernon Wheeler- of AWOL. He was my father only because he was documented on my birth certificate. Psychologically, I had no relief from my feelings of abandonment over time. Each State Hospital visit as if my body was trapped in a bulky waterproof diving suit, lead boots dragging legs backward each ascent, lungs gasping air leaking cracked air tube, detachable helmet face mask muffling my screams. I was just as abandoned as an adult as I was as a child.

The criminal justice track involved coordinated internships in juvenile correctional settings and core courses in clinical casework, human development, community organization, research and statistics, social policy, group work, and welfare economics. In 1964 most M.S.W. students aspired to work in traditional private Family Services settings and County Child Welfare agencies. The general clinical model was coercive court-ordered rehabilitation vis-à-vis therapist/client or group interview and evaluation to address socially dysfunctional behavior stemming from child abuse, drug, and alcohol addiction, loss of self-esteem, and victim of violence. The objective was to identify the client's individual and environmental factors affecting his or her social dysfunction. Therapy techniques reflected the School of Psychology of psychiatrists or trained psychiatric social workers teaching courses in Psychoanalysis, Behaviorism, Gestalt, and Holistic treatment. Regarding community organization specialization, the dominant model was mobilizing resources from voluntary agencies such as *United Way, Red Cross,* and philanthropic groups. Sadly, course reading lists rarely included James Agee, *Let Us Now Praise Famous Men*, Saul Alinsky, *Reveille for Radical,* John Steinbeck, *The Grapes of Wrath*, C. Wright Mills, *The Power Elite*, Michael Harrington, *The Other America, or John L. Lewis* American leader of organized labor who served as president of the United Mine Workers of America from 1920 to 1960. But, when it came to mentors in casework, internship, and community organization, I drew a Royal Flush.

For research and statistics and my master's thesis advisor, John Behling, M.S.W. Ph.D. Sociology took me under his wing. John started as a social worker at Columbus Southside Settlement House in the early 1950s. He rejected the traditional casework intervention model that leaned more toward helping the vulnerable adjust to poverty or pathology of living conditions than attacking the source of their oppression. He advocated radical social work and was an early opponent of the Vietnam War.

W. Hugh Missildine, M.D. was the director of the Children's Mental Health Center in Columbus, Ohio for nine years before he was an Assistant Professor of psychiatry at the Ohio State University. I had the double fortune of taking his class and working under him as an intern on a special case involving a 14-year-old male committed to Ohio Youth Commission's most secure institution, the Training Institute of Central Ohio (TICO) for violent juvenile offenders.

A Case Study Of Patricide & Matricide

I sat alone in the social worker's office at TICO, studying the *Polaroid* images of double homicide: the mutilated face of a middle-aged white adult male shot at close range by a 12-gauge shotgun; the body sprawled on a barn concrete floor in a pool of blood. The only witnesses were a dozen registered Holsteins attached to state-of-art four-teat *De Laval* milking machines. Another photo showed a small woman spread-eagled face down in another red pool on a Spanish tile kitchen floor. She was shot in the back, hand clutching a muffin pan. According to the Sheriff's report:

> *14-year-old male youth Danny Mauger was found sitting and rocking on the kitchen floor, cradling a pump shotgun next to her body. The forensic analysis speculated that the boy shot his father first in the barn a hundred yards away from the house because his mother was hearing impaired and his dad would have heard the shot. Neighbors passing in a car heard a gunshot and called the sheriff. Within hours of arrest, Danny's family relatives came to his defense, pooling money to hire a lawyer. A neighbor who had done some odd jobs at the Mauger farm reported the old man, had a reputation for treating the boy like a slave, and was chintzy paying hired help. ` Whenever I've seen the boy with his dad in town, the kid's behavior reminded me of an abused puppy. He'd flinch, and recoil at the sound of his father's voice and mean stare.`*

Dr. Missildine outlined his diagnosis and treatment plan for Danny Mauger in the context of his lectures and the text of his world-renowned classic book, *Your Inner Child of the Past*. His goal was to help free Danny from the enslavement of three pathogenic or troublemaking parental attitudes: (1) **perfectionism** *created in the child by the parent's withholding acceptance of the child until his behavior is more mature than he can comfortably achieve at the time; (2)* **over coercion** *expressed by the parent who constantly directs, supervises, redirects the child with an endless stream of anxious reminders and directions;(3*) **punitiveness** *widely accepted in our culture as something necessary for the disciplining and training of a child.*

Danny was escorted from class by a tall Black Youth Leader dressed in civilian clothes. A *staff* photo ID was clipped on his shirt. Danny wore pressed jeans and a denim shirt with a *student* photo ID. This was Danny's second day at *TICO,* following two weeks of psychological and educational tests at *OYC's Juvenile Diagnostic Center* next door.

To begin on equal footing, I walked around the desk, opened the door, and extended my hand to a slim stooped figure, his dark brown eyes scanning naked institutional walls before shaking my hand. It must have been a constant reminder to him of where he was. His dark hair had recovered its part since JDC's intake burr cut, face pale from confinement, a few teenage zits but not unsightly blemishes. He slouched in his chair, folding his arms. I introduced myself as a social worker intern working with Dr. Missildine who evaluated him for the court. I told him I read his file but needed to hear what happened in his own words.

I said, "Danny, our mission is to help you process what brought you here and what is required to convince the court Danny Mauger is no longer a threat to society." I paused, waiting for him to establish eye contact, unfold his arms, and sit up in the chair. "Danny, do you remember what happened the morning you were arrested? Did your dad or mom do something that upset you?"

Danny sat motionless, mute, eyes cast downward. Aware of his father's dictatorial control over his son's life as described by Danny's neighbors, and kin and cited in Missildine's psychiatric evaluation, I opted for equal silence to convey mutual respect toward Danny's feelings and freedom of choice not to verbally respond to an authority figure.

Ten minutes passed before Danny broke the ice, saying in a contemptuous tone. "He.... caught me popping a pimple on the bathroom mirror the day before. Said I acted like white trash. Threw the toothpaste tube at me for squeezing it in the middle. Bitched about me using too much toilet paper he rationed cause he is so chintzy. Gave me another strapping before my morning chores." Danny continued, speaking rapidly, pausing to take deep breaths. He described how he sneaked his father's shotgun out of his parent's bedroom closet the night before the crime, loaded it in the dark, stayed awake all night, smeared toothpaste on the bathroom mirror, unraveled a roll of toilet paper and toilet-papered, then waited behind a stall in the barn. Knew if he shot his mother first his father would hear and react to the gunshot. Whereas his mother was hearing impaired and not warned. He watched his father rush into the barn and retrieve the leather strap used repeatedly on his young son's bare buttocks and legs from a hook whenever he was dissatisfied with Danny's response. That morning, his father was on fire, looking for him after seeing the bathroom and Danny's empty bed. Danny stepped out of a stall, aimed the weapon at his father's head, and fired point-blank.

"What about your mother?"

"At home, she always left the room when he got red-faced and went berserk".

"What happened after you shot him?"

Danny's eyes welled and turned inward.

After a long pause, I asked, "Do you have any hobbies?"

"Collected rocks with my grandfather. He used to take me rock hunting."

"That's really interesting."

"He died last year."

"I am sorry about your grandfather."

"He was my mom's dad. Didn't get along with my father."

"Why?"

"Rock hunting interfered with my chores."

"Danny, I appreciate your openness. Unlike many boys in places like TICO, you have an unprecedented family support system to help you through this crisis. Your Aunt Betty Johnson and Grandmother Marybelle are very empathic to your current situation and committed to finding the best path forward."

The treatment plan included two weekly sessions with Danny and bi-weekly conferences with TICO staff chaired by Dr. Missildine. I also coordinate family visitations.

Reflecting on my *inner child of the past,* Missildine's paradigm was an insightful framework for understanding the effects of *pathogenic parental attitudes* on adolescent and adult behavior, particularly as it pertained to the dynamics triggering Danny's violent act. Introspectively, my mentor's paradigm further illustrated the contrasting advantages bestowed on a past-around orphan of the Carney clan. The sessions tracing Danny's footsteps to the crime revealed the psychological damage of his father's excessive *perfectionism, over coercion,* extreme *punitiveness* regarding his son's farm chores, relentless bullying, and public belittling, which is anathema to developing a young boy's self-confidence. Reflecting on my early childhood freedom, love, and support at the Carney's homestead and attachment to Grandpa Carney's hand-carved toys, and later models left behind at the Watson's day commencing California exile, I sympathized with Danny's need to reconnect to comforting mementos of his past life with his grandfather.

I opened the latch gate of a white picket fence enclosing the Mauger's country house and surveyed a picture-perfect landscape of a farm in the Ohio heartland: barbed wire fence surrounding 300 acres, several centuries-old oak and pine, a silo, a new John Deere tractor and an all-terrain vehicle parked out of the elements inside a modern hanger-shape corrugated metal structure storing farm supplies: feed sacks neatly stacked, milk pails lined in a row, pitchforks and shovels secured on wall hooks, water hose coiled and dirt floor. A recently painted red barn designed to accommodate two dozen milk cows, a footpath leading to a pond. The deceased dairy farmer's livestock had been transferred to a neighbor's farm. The farm equipment appeared clean and organized. I removed the lens cover of the 35 mm *Nikon* camera and entered the main barn. The musty smell of three months of dust, scattered hay and manure missed in the cleanup of the crime scene permeated the interior of the barn. A few empty milk cans and coiled vacuum hoses were left on the concrete floor of the milking stations. I snapped a picture of the whipping strap hanging on a hook in the natural morning light of the only window, preserving the evidence of child abuse.

I moved on to the main house. Except for a faint covering of dust, the furniture was in place and looked elegant. The decorator of the house favored a mix of Ethan Allen, antiques, thick neutral carpet, and floral curtains. The kitchen was befitting a chef's dream: built-in dishwasher, *True TS-49-HC* solid stainless-steel refrigerator, custom-made wood cabinets, giant cutting table, double oven, and gas stove. It had a walk-in pantry filled with goods and supplies that could survive a siege.

However, Danny's bedroom appeared Spartan. No signs of stereo, teenage posters of superstars or models. Only books were school textbooks, a hard-back *1st edition Introduction to Geology by Branson and Tarr1935,* and glass-encased cabinets displaying trays of rock collections. Each rock was meticulously labeled. There were chunks of blue anhydrite, translucent calcite, sharp pieces of flint, polished quartz, and pale brass-yellow pyrite or *fool's gold*. No knick-knack items, toys, or ball gloves on wall shelves. I found a backpack containing a small hammer and a topo map of eastern Ohio. The master bedroom reflected a woman's tastes--still life oil paintings, *King James Bible*, artificial flowers, an antiqued doll collection, a floral quilt bedspread, and a vanity table. Only one photograph adorned the walls-- the victims' wedding picture. In contrast, Mr. Mauger's knotty pine den displayed several photographs of him proudly accepting awards over the years from the *Ohio Dairy Farmers Association*. If I hadn't seen Billy's bedroom myself, I would be hard-pressed to believe a teenager lived in this house. Before leaving, I gathered some of Billy's things in the backpack.

At our next session, Danny stared wide-eyed at a small arrowhead-shaped stone resting in the silk lining of a rock collection case, His fingers nimbly extracted it, and held it to the light. Its surface glistened with rust-colored streaks and iron-gray scales. He lay it in his palm, exposing a fossilized fern on each interior side. His fingertips lightly touched a thread-size stem, tiny leaves and veins sculpted over eons like an artist's relief.

"Did your grandfather help you with this collection?"

Danny nodded.

"When was the last time you and he went rock hunting?"

He tilted toward me, eyes locked on the fossilized fern cupped in his hand.

"Two birthdays ago."

"Where?"

"Flint Ridge in Muskingum County."

"Must have been very sad after he passed."

He pressed his lips and blinked repeatedly. His shoulders sagged.

I paused, waiting for him to recover, and reached into the backpack. "Danny, I brought you a book, *Minerals of Ohio,* I thought you'd like, and other rock books I found in your room. TICO residents can't possess rocks, but books are OK."

Eight months later, the State prosecutor argued against Danny's mother's sister's petition to be declared Danny's guardian, change his legal surname, and terminate OYC's commitment. According to the *Ohio Revised Code,* the judge has considerable discretion in juvenile cases. Thanks to Dr. Missildine's psychiatric evaluation the original prosecutor was overruled in having the court certify Danny as an adult and sent to Ohio State Reformatory, Mansfield.

The new prosecutor cross-examined Dr. Missildine, reminding the court of the juvenile defendant's premeditation and extensive planning involved in carrying out the heinous act: shooting his father first, point-blank, then taking advantage of his mother's hearing impairment by shooting her in the kitchen. Given the court ruled out grounds for insanity or incompetency, releasing from *TICO* was a threat to public safety.

Dr. Missildine countered that we believed there were serious mitigating circumstances. The more we investigated the circumstances, and followed his progress at *TICO*, the more the treatment team was convinced that Danny Mauger acted out of self-defense. His father enslaved him with chores and beat him with a leather strap as a matter of course while his mother stood by or ignored the abuse acting as an enabler.

The judge called both sides into his chambers after the noon recess. He instructed our lawyer to submit the name of a guardian to administer Danny's substantial estate and change his legal name as we outlined in our petition.

OSU Campus Uprising

Beginning March 1964, *Operation Rolling Thunder* B-52s dropped tons of bombs in North Vietnam, while Navy F-105s splattered South Vietnam villages *with Hellfire* napalm, losing fifteen to thirty planes per thousand sorties. In the spring of 1965, The Ohio State University campus was teeming with anti-Vietnam war demonstrators and students determined to overturn the OSU Board of Trustees' ban on communist guest speakers.

Reacting to uncensored daily TV Vietnam war images, news of spiking US casualties, ARVN defeats, and the call by military leaders to raise draft levels and activate reserves and National Guard, I went with my academic advisor and mentor John Behling to the first *Vietnam War Teach-in*, March 24-25, 1965, at University of Michigan, Ann Arbor. Over 3500 people opposing the war came from across the country to participate in a series of discussions and debates led by academic experts in community organizations like Jack Rothman, M.S.W, Ph.D., Social Psychology, and student leader Carl Oglesby, President, Students for a Democratic Society (SDS). Speakers presented strategies of achieving participatory democracy and equal justice; principles opposed to America's unilateral intervention in Vietnam's civil war; a country with a thousand-year history of driving out foreign invaders, e.g., Japanese, French. Attendees dove into anti-Vietnam war articles published in underground newspapers e.g., *Los Angeles Free Press*, *OZ Magazine,* and Bernard Fall's historical study of Indochina, 1945-65, *Street Without Joy,* the moniker of Road 1, 20-mile, 300-yard wide north-south artery from Hue to Quang-Tri along Vietnam coast bordered by a labyrinth of fortified villages with a history of black-clad Viet-Minh and French carnage.

The following month, I marched on High Street near campus with war protestors pelted with eggs and epithets by counter-demonstrators. On April 28, I joined an estimated 300 students, sitting outside OSU President Fawcett's office, demanding that the Administration change the Guest Speaker Rule which essentially banned Dr. Herbert Aptheker, national director of the *American Institute for Marxist Studies* from speaking on campus. Some members of the unofficial student organization *Free Speech Front* of which I was a member stayed all night. The rumblings of discontent over the Speakers' Rule continued intermittently until October 1965 when Ohio Governor James Rhodes intervened to stop the negative publicity of campus disruptions provoked by the OSU Board of Trustees' free speech rule banning controversial student guests' speakers like Aptheker; a rule inconsistent with other Ohio Universities. Rhodes appointed James Shocknessy a conservative Democrat, to one of the positions that had become open. His vote represented a switch in favor of effectively abolishing the rule.

To get a head start on the Social Work Master's thesis requirement, I completed a review of the literature and surveyed union and non-union welfare caseworkers in Gary Indiana, and Columbus respectively. The research design involved developing, administering, and analyzing an attitudinal scale measuring the *egalitarian orientation* of employees at unionized vs non-union welfare agencies.

War Games

In August my attention turned to carrying out duties as 2nd Lieutenant Medical Services Officer (MSC) in the Headquarters Battalion of Ohio National Guard, responsible for organizing logistic support of medical services field operations for two-week war games at Camp Grayling, Michigan, the nation's largest National Guard training center. I was the lowest rank non-physician or dental officer in the HQ unit, overseeing transporting via highway convoy, medical equipment, and supplies 402 miles from Columbus to Camp Grayling. Moreover, I was the principal staff officer charged with setting up medical field operations within the first 24 hours for thousands of guardsmen and regular army personnel participating in annual war games encompassing a 147,000- acre landscape of dense white pine. The area included a downhill skiing area for 11,000 feet elevation of mountainous terrain with a ski lift and 1,920-acre Lake Margarethe.

The most difficult task was meeting the 24-hour timeline. Provided a jeep and driver, I carried out duties as a staff officer representing the HQ Colonel and the medical brass. Another task was teaching new medical officers commissioned as captains how to salute. Our unit shrink admitted that he was allergic to the sight of blood. The Battalion Commander roused us out of bed at 0500, blasting the OSU Marching Band's fight song over the PA system, followed by reveille. My most anxious moment occurred the second morning when ordered to report to the brass overseeing the war games. As I entered a giant tent, I could tell by the shoulder patches it was a mixed group of National Guard and regular army senior officers. I was immediately waved to the front of the room by a tall, chiseled face, steel-eyed, regular army *three-star* general dressed in starched fatigues, a *Combat Infantry Badge* and *U. S, Airborne* parachute symbol planted on his chest, and spit-polished paratrooper boots with white laces. Standing erect side-by-side, facing a topo map of Camp Grayling's training site on the wall, he pointed a swagger stick at a remote spot in the middle of the map.

Shifting his eyes from the swagger stick stabbing symbol of raised terrain to the wide eyes of the lowest rank officer in the room, he said, "Lieutenant Wheeler, you just received a radio communication from a rifle platoon leader that one of his men crashed a jeep into a tree and lays unconscious in the weeds."

I smartly stepped up to the map, eyes fixed on the tip of the swagger tip to contact the medivac unit and order them lance and medic with resuscitator ASAP, then I'd, instruct them to scramble to the corresponding coordinates and life flight injured soldier to the base hospital if medic so determined.

Following a brief pause, the *three-star* said, "Thank you, lieutenant. Give my regards to Colonel Good."

My routine day consisted of being driven around the woods in a jeep or ambulance checking the medical field tent operations for supplies and responding to emergency calls in remote areas. We got flagged off-road by a dehydrated guardsman who forgot to fill the canteen before a long march in the woods. To break the boredom, I'd watch *Sherman Tanks* drag-racing in rolling hills, visit OCS classmates in the artillery brigade; tracking the arc of 105 mm howitzer projectiles like watching BB shots from a *Daisy rifle.* One morning, I filled in for a sick rifle range pit officer, a dangerous "shit detail" because unlike a carnival shooting range, whereby a *Remington* pump rifle firing 22 (short) at the mechanical moving row of ducks, guardsmen shoot U.S. Army M1 *Garand* 7.62mm caliber semi-automatic bullet that obliterates a toy duck and continues through the barrier to carousel rider. The Army rifle range was a manual operation in which a dozen low-rank weekend warriors stood in a four ½ foot deep, 20-yard-long ditch, each manning a poster-size paper target raised for prone shooters 74 yards away, then lowered for pit markers to score. The Pit Officer quickly learned that this was a dangerous drill. Notwithstanding wearing a steel pot that was not bulletproof, a micro-second exposure risked a fatal wound from a lethal round piercing a paper target or ricocheting off the ground like swarming bees attacking hunkered men. The only remedy, Pit-Officer**,** madly racing the pit line, crouched down, shouting, "**Heads Down,** motherfuckers," swagger stick hitting helmet tops of soldiers violating rifle range pit protocol.

Meanwhile, according to newspaper reports, the average monthly number of the U.S. killed in America's real war ten thousand miles from their homeland soared from 18 in 1964 to 160 monthly actions in the first six months of 1965. In July 1965, President Johnson announced he would increase total American troops to 175,000. During fake battles, medical staff in the field closed the tent flaps, made GI cocktails, mixed medicinal alcohol, and gin, drank contraband beer, played poker, and listened to red team (regular army) invaders firing blanks at retreating part-time warrior foes. A lucky straggler stumbled into our tent, treated with medicinal spirits, and sacked out on a gurney reading an underground newspaper.

During fake war intermissions, NG and RA officers commingled into Camp Grayling's officers' club. Many Regular Army (RA) officers had recently completed combat tours in South Vietnam. I observed only one NG Black officer, a 1st lieutenant sitting alone at the bar, nursing a bottle of *Coors*. Cognizant that Blacks were underrepresented in Ohio's Army National Guard officer corps, I sat next to him and ordered the same. Shaking hands, we exchanged names, civilian occupations, and hometowns. He said his mother named him Cody for the town where he was born. He was a high school history teacher in Cleveland. He completed 6 months of Private E-1 active duty after graduating from Bowling Green, went on to OCS, and now was a platoon leader in an infantry unit. I told him I was an OSU social work grad student hoping to work with juveniles. And the most fun I've had at Grayling was getting tanked on beer and riding the local ski lift with the medical battalion dentist.

The Camp Grayling Officers Club ritual was eating popcorn, smoking *Owl* cigars, and drinking two-bit PX beer. Tonight, the black and white TV screens showed a young eager reporter wearing a Texas A &M ball cap, clean GI fatigues, *Abercrombie & Fitch* photographer vest, and civilian hiking boots. He was approaching an unshaven white Marine officer in tattered fatigues, wrestling out of a shoulder harness supporting a field pack, ammo belt laden with smoke and fragmentation grenades, two canteens, first aid kit pouch, holstered .45 semi-automatic, and KAY-BAR in the leather sheath. The TV camera panned a small group of dirty face jarheads sprawled in tall grass and zoomed on their leader's jettisoned helmet, M-14 rifle, equipment, and drawn face. Then the camera showed the reporter raising a hand and holding a tape recorder to the marine's face.

The reporter said, "This is Marine Corps Lieutenant Norman Wallace. He just returned from a reconnaissance mission somewhere in South Vietnam. Tell us, lieutenant, how many days was the patrol?

"Classified," he said as if to deter a follow-up question from an unauthorized war correspondent.

"Y'all kill any communists?"

"No," he said, curtly.

"Aren't you disappointed?"

"Negative. I achieved my mission."

A reporter with a puzzled expression, said, "Mission?"

"All my men– returned alive."

"Jesus, that's one jarhead lieutenant who won't make captain," said the NG lieutenant from Cleveland.

1965-1970

Watts

Camp Grayling Officers Club, Breaking News, *Watts Riot Day 4:* TV images of the deadly uprising in predominantly Black south-central L.A. neighborhood being suppressed by an army of National Guardsmen with fixed bayonets, white cops with shotguns, and police dogs. I related to Cody that I was aware of L.A.P.D.'s negative reputation with Watts. Like in Cleveland and Columbus, Blacks were corralled into ghettos by racist real-estate policies and poverty. As asserted by Michael Harrington, Blacks were invisible, out of sight, out of mind to whites in suburbia. I confessed my first serious reaction against racism was quitting an all-white college fraternity, *Sigma Pi*. Bigfuckingdeal. My epiphany was seeing TV news of white rabid segregationists torching a Greyhound bus that brought civil rights workers to Jackson, Mississippi, followed by police mass arrests, beatings, and jailing of civil rights advocates. This inspired me to hitchhike to Jackson's Greyhound bus station and sit in the *color-only* restaurant there. Before I could finish a Coke, fucking crackers dragged my *Yankee body* to a waiting police cruiser and two of Jackson's finest that drove me out of town to an eerie swamp. Fearing for my life, I produced my white privilege race card. Gave a spiel that was more bluff than real. Said my family will send Federal Marshals searching for me. Truth be told, my escape north was aided and abetted by a courageous old Black junk collector hiding me in a smoking pickup.

August 15, 1965, brought the notorious Watts Riots Day 5: civilian Body Count: 34; Cop Body Count:2.

Camp Grayling officer's Club television screens flashing surreal images of American flag-draped coffins unloading at Dover Air Base, U.S. Marines burning huts of rice villagers, grunts humping dense jungle searching for the invisible enemy, gunships strafing sampans in winding canals. Cody motions to the bartender for more *Coors*, his dark eyes tracking the reflection of fading cigar rings in the wall mirror.

I told Cody about the Vietnam Teach-in at the University of Michigan, and what I learned from reading Bernard Fall's *Street Without Joy*, the irony of America supporting South Vietnam's corrupt regime rather than Ho Chi Minh, the popular North Vietnam hero that quotes the U.*S. Declaration of Independence*, and whose primary military and political objectives are establishing an independent united Vietnam. How, in my humble view, Ho Chi Minh's leadership was more analogous to Lincoln in our Civil War fighting to preserve the union and abolish slavery than South Vietnam's Diem oppressive authoritarian rule over South Vietnamese peasants and Buddhist priests.

I alluded to my conversation with a two-tour medevac helicopter commander. A Gene vehemently opposed U.S. intervention in Vietnam. He said, "American equipped ARVAN troops are afraid to engage Viet-Minh or Vietcong in South Vietnam and their U.S. trained medevac pilots who refused to rescue their own wounded. Consequently, our medevac crews answered the call and raced to the killing zone identified by the fading purple haze rising from the smoke bombs of survivors. All we found were stripped ARVAN bodies, dog tags, and bone chips of American advisors. Never read this shit in the after-action reports or the *New York Times* did ya?"

"No," I responded.

Our CO announced that the war games were to be suspended two days early. We were ordered to strike our tents, load equipment, and be ready to return to Fort Hayes by convoy at 1300. The scuttlebutt was that the Watts Riot was making the buckeye politicians nervous: as if the Ohio National Guard may be needed at home. Another rumor we had to grapple with: As a precaution to deter white "farm" part-time soldiers rushing home to deter or quell a race riot in the city, the M-1 firing pins will be removed. No Shit. Who'd believe Ohio Guardsmen would shoot civilian protestors?

June 1966. I was offered a Ph.D. fellowship in Social Work at the University of Pittsburgh. I rode the elevator to the 40th floor of the *Tower of Learning* and scanned the looming sky, suddenly vertigo overcame me, and I fell back to earth. Meanwhile, my wife mulled over accepting a teaching job at a fenced-in school on a cobbled street in Mckees Rocks. Eventually lured by California teaching and MSW jobs posted in professional journals, we loaded our belongings in a U-Haul trailer hitched to VW, jumped into a 63` Porsche convertible we paid $1900, borrowed from Teachers Credit Union, and left our dog, Braun with Carol's folks until we got settled in Seal Beach.

Special Services for Groups, a non-profit social services agency, paid me $600 a month funded by a summer grant to divert Watts children at risk from L.A. County's juvenile justice system. I reported to a middle-aged Black woman at an office at 103rd and Compton, the epicenter of violence the previous year. Her modus operandi was a passive Family Services clinical model whereby middle-class clients voluntarily walk in seeking family counseling. My challenge was to reach out to community leaders to identify children at risk of falling into the cracks of child welfare or the juvenile justice system. In my experience, unlike law enforcement, Social Workers were not anathema in poor communities, regardless of race or ethnicity. Ideally, we served the casualties of racism and unfettered capitalism by identifying and ameliorating institutional bureaucratic impediments to delivering basic human needs such as shelter, safety, education, employment, childcare, and health services to poor families. Thanks to L.B.J. 's 1964 *Office of Economic Opportunity Act,* some of these issues were being addressed post-Watts riot due to the rise of *Black Power* inspired by Adam Clayton Powell, Stokely Carmichael, and the *Student Nonviolent Coordinating Committee (SNCC)*. In the summer of 1966, the country was reeling from the civil rights, anti-Vietnam War, Berkeley Free Speech movements, the shooting of James Meredith in Mississippi, rise of alternative press and counterculture. L.A. government officials and leaders did not want a repeat of the fire, blood, and death of August, 65, or the violence depicted in the South.

My primary mission was to divert children at the most vulnerable stage of human development from pathological neglect endemic in poverty. They experienced neglect that was exacerbated in the summers due to the void or absence in poor communities of educational, recreational, and cultural enrichment programs available in working-class and white middle-class communities. In contrast to Columbus Eastend's Roosevelt Park, vacant lots, free Knothole Gang baseball games, freedom of mobility to the Ocean beaches, free fishing piers, amusement parks, and cheap movies in Long Beach, until the War on Poverty, the children of Watts were ruled by fascist cops, social isolation, and Lord of the Flies mentality. In the summer of 1966, Watts experienced a temporary peace with L.A.P.D. Community Action Funds created jobs and hope for younger adults and impoverished families.

I attended a church gathering of Watts residents discussing new grants for a tutoring program for students failing elementary school. It offered a two-dollar-a-day financial incentive to poor students to attend class and a bonus for demonstrating academic progress. The grant covered a tutor, textbooks, flashcards, and supplies. Additional grant money paid working-age boys on probation and parole to clear away riot debris and renovate damaged houses. The *Black Power* movement also attracted troubled teens to participate in life-affirming art, educational and constructive projects; amenities long denied to minority communities. However, mothers and social workers expressed concern that many pre-teen boys did not respond to the War on Poverty social programs directed at older youth.

Other than uniformed postmen, whites in civilian clothes making home visits during the day in minority neighborhoods were generally not at peril because they were social workers trying to help. However, searching for children wandering the streets or hanging out in abandoned houses, or knocking on doors after dark was another matter. Seen as a stranger, one could be mugged and robbed, or have their car stolen. Parents were anxious over the county's standard procedure of taking minors found wandering alone on the streets directly to Maclaren Juvenile Hall and letting Child Welfare intervene. After meetings with stakeholders including Child Welfare, law enforcement, non-profit agencies, and Black Power leaders, I concentrated on establishing ground rules and strategies to attract young boys missing in action to activities tailored to their needs. Following several nighttime ride-alongs with plain-clothed Black juvenile officers searching for preadolescent boys rumored hanging out in alleys and abandoned buildings, I negotiated an alternative strategy of taking stray children home and follow-up scheduled home visits. Fearful parents associated surprise home visits with child welfare abuse or neglect investigations. The previous coercive intervention model is often triggered by teachers, hospital emergency medical evidence e.g., x-rays, photos of torture, malnutrition, and sexual abuse. My target population was preadolescent normal rambunctious boys of color seeking mastery and self-discovery in a post-racial war environment that continued to marginalize them. Indeed, in 1965 Watts was the Black Warsaw of Los Angeles County. In 1966 the County Establishment/Watts community appeared committed to summer peace. Unfortunately, the country was not.

There I was, a casually dressed 26-year-old white guy sitting on a footstool, in a modest dwelling a few blocks away from last year's Watts carnage. The living room was furnished with three generations of furniture and corresponding photographs of family history, Rev. Martin Luther King Jr., John F. Kennedy, and two individual color portraits of an identical young Black man: one wearing a graduation robe, waving a high school diploma, the other, stern face, in U.S. Army private enlisted Class A uniform, probably taken after boot camp graduation. The young GI's picture stabbed me with guilt over my white privilege college deferment allowing me to dodge the Vietnam war draft. Only one close friend, the ex-Golden Bear Trent Lamb, enlisted in the Air Force and served in South Vietnam. A few others were commissioned officers in the medical corps, pilots, and lawyers. Meanwhile, I continued rounding up stray children in Southcentral L.A., joining protest marches, painting peace signs in public places, mailing letters to congressmen, and procrastinating transferring to California National Guard: fucking facile. Mid-July. I was looking at the olive skin face of a middle-aged, grandmother in a coral house dress, overweight body sinking in a worn fabric couch. Her shoulder tilted into the skinny eight-year-old chocolate skin male child in a clean white cotton t-shirt, baggy jeans, and unlaced worn sneakers barely touching the faded carpet floor.

In a quiet voice, she said, "Harold, say hello to Mr..."

"Wheeler," I interjected, extending a hand he ignored. "Harold, I am a social worker starting a program here for boys your age..."

Personal safety came up during the initial community meeting. A Black *California Youth Authority* parolee volunteered to ride along in searches of abandoned buildings. With his help, I became the pied piper for Watts wanderlust boys. Dark skin Black and Hispanic youngsters tagging along that had never seen the ocean less than 20 miles away, made sandcastles, rode paddleboards, pier fished, played touch football in a public park, and listened to their voice on a tape recorder describing field trips to museums and libraries. As the word spread, my four-passenger field trips in VW beetle resorted to shifts until the transportation problem was resolved by acquiring a surplus U.S. Marine recruiting station wagon which raised some eyebrows driving around town. These boys lived in broken homes in an isolated segregated impoverished community with infrastructure comparable to ruins of war. The principal caretakers were single mothers, grandparents, aunts, or older sisters. Fathers were in jail or absent by punitive welfare policies that cut child allowance if the I fathers lived at the home. These racist policies broke up far too many families. Older male siblings lacking vocational training or education survived in the criminal street economy of drugs, and gang violence, or became cannon fodder for the class-based military draft. In Watts, I perceived myself as part Peace Corpsman, community organizer, and clinical social worker.

Using social group work skills, I scheduled field trips and led weekly group meetings composed of Black and Hispanic boys ranging from ages seven to ten. I linked boys with poor school histories to Watts' tutoring programs. A difficult daily challenge was addressing intra-racial slurs among Blacks in competitive games and altercations like I had witnessed at basketball games at TICO. Clinicians attribute Blacks' use of the derogatory slur, *nigger*, to fellow Blacks as the expression of self-hate. When race talk got out of hand, I suspended the activity and held group meetings, stressing its destructiveness. Related that race talk incites violence, pointing out it was rare to hear Catholics calling each other *mackerel-snapper,* Italians-*dago*, Mexicans-*wetback*, Native Americans-*redskin*, Jews-*kike*, Asians-*slant-eyes or slopes.*

The Children at Risk federal grant ended the first week of school. The white's notion that Black welfare families fail to instill pride in themselves and their children was blatantly prejudiced and false. Notwithstanding severe financial hardship, Watts' mothers were no less meticulous than their affluent counterparts in dressing children in their Sunday best for the first day of school.

During this period, I learned that only one California National Guard non-physician vacancy Medical Branch junior officer existed in Los Angeles and Orange Counties. In this case, it was an attached slot at an Artillery Brigade rather than a Medical Battalion. The protocol required an initial interview and approval of the unit's commanding officer, in an office befitting a warrior. Standing erect in street clothes and custom-made leather sandals, with six weeks of hair growth, combed by wind, facing a full bird colonel with an oversize shaved head, hunched over a mammoth desk laden with symbols of war, I said, "Second lieutenant Gerald Wheeler reporting, sir, " snapping a salute.

"Serial Number?"

Taken aback, reciting a number long forgotten, I reached into my hip pocket, fumbling my wallet to retrieve a military ID, reading the number aloud.

"What are your career goals in the California State Guard, lieutenant?"

"Sir, complete my military obligation. Two years, I think."

"Attending Senior officer training required of Captains?"

"Complete my duty in the guard, Sir."

I received no word for a month, I called the NG HQ and M.D. that referred me to the bonehead Colonel who arbitrarily rejected the transfer request. A rejection triggered a letter addressed to Gerald R. Wheeler, re-classified 1-A. Subsequently, I sent a protest letter to Senators, J. William Fulbright, Stephen Young, and Wayne Morris, accusing a California National Guard of discrimination against a bona fide Commissioned Officer. Soon thereafter the Army reserve sent a letter informing me that I was transferred to "active reserve officer status until further notice," which expired years later when I happily signed the honorable discharge papers.

Camp Vernon J. Kilpatrick A.K.A., Krazy Kamp

During mid-September, 66, I stood on the 16th floor, gazing out a window overlooking the vital organs of Pacific Coast's Gotham City. Luckily, a woman's soft voice broke the silence. I was no longer invisible.

"Mr. Osterman will see you now," she said, opening the door, and gesturing to proceed.

I stepped forward, facing a big Italian-looking man who had a partially bulbous red nose like W. C. Fields, massive head and shoulders hunched over a stack of papers on an executive desk flanked by Gold, Blue Los Angeles County flag, California Grizzly Bear flag and Old Glory. A bright floral tie dangled loosely from a strangling collar of a pale blue oxford shirt. His left thumb and index finger held a half-smoked fat, black cigar, its growing ash precariously close to falling on the reading material. I welcomed the aroma, eying a varnished wooden plaque on the wall behind him, boldly branded as follows: ***Sam Ostroff Director Los Angeles County Probation Department Juvenile Institutions.***

The County official looked up, and deftly tapped the cigar, its ash landing in a solid green color glass pipe ashtray. He had the take-charge wrinkled face of a mafia don; bushy brows arcing dark eyes signaling his brain of my sudden presence and purpose. His thick dark hair was disheveled, probably by self-massaging pondering problems flooding the in-basket on the desk. I anticipated being interviewed by a non describe civil servant at *MacLaren Hall Children's Center* that advertised an MSW counselor position in the *National Association Social Worker (NASW) Newsletter.*

Rising behind the desk, in shirtsleeves, his freehand holding my resume, the other, a dying cigar, he said he was impressed by my having attended an atypical social work graduate program that offered a concentration in juvenile corrections, my TICO internship, Dr. Missildine's and Professor Behling's glaring letters of recommendation. He was familiar with *Special Services for Groups* outreach programs, and lastly, my Army training. He candidly described the L.A. County's Probation Department's juvenile correctional system which he oversees: *MacLaren Hall,* L.A. County's infamous overcrowded detention center, and the county's dozen forestry camps designed for older male delinquents that work clearing brush for firebreaks to mitigate wildfire damage. Boys that threaten public safety i.e., homicide, rape, and gang violence go directly to *California Youth Authority.* He said the population includes school dropouts, runaways, incorrigibles, kids into petty crimes, drug abuse, and multiple probation violations. These forestry camps have no full-time school, except volunteer tutors. He continued. In the early 1960s, the enlightened County Board of Supervisors funded Camp Kilpatrick, a special residential treatment and education program for adolescents facing a new set of social problems, drug abuse etcetera. Behavior that was attributable to problems at home, low self-esteem, poor peer relationships, and emotional neglect. In the home or community, their anger has the effect of a tidal wave hitting an insecure beachhead, moving with great thrust, and destroying everything in its path. They demand attention 24/7, compelling staff to work through mutual conflicts, depression, and anxiety related to separation from family, while at the same time, immersed in an intensive unpredictable camp environment which can change from calm and chaos in a wink. To avert catastrophe, wayward boys and exhausted staff must be resourceful, cooperative, empathic, and creative.

Presently, he noted, Camp Kilpatrick was staffed with twenty-one Deputy Probation Officers. All had a college degree or more. He said, "The individual casework is damn good, but DPOs need a treatment model that will address the stress and fatigue that comes from engaging in ongoing conflict resolution with 90 rambunctious boys at war with society and themselves." After this, he paused, "Gerald, according to your resume, I think you got the right toolbox. I'll swear you in now, issue you a Senior Deputy Probation Officer gold shield, three-step-paygrade."

I quickly agreed.

Strapped in the leather seat, left-hand iron-grip on steering wheel, heart revving as fast as 63` Porsche (top-down) Cabriolet, air cool 356 engine, *Blaupunkt radio blasting rock & roll*, *Pirelli* tires spitting gravel, right hand grasping vibrating gear stick downshifting narrow mountain switchback curve. Vertigo be damned. Inhaling salt air, I scanned a mountain landscape of dark-green thickets, thriving black sage, tall pine, sprawling live oak, and patches of burnt stumps and log remains of lightning strikes. I stared at canyons below filled with pockets of purple-colored tubular wildflowers resembling a duster used by Grandma Carney to clean her delicate figurines. I squinted at rocks and granite cliffs glinting pink, blue, and orange bands of the bright noon sun, golden waves of wild grass, and tiers of green trees. The only incongruous features were swaths of firebreaks wide as a young boy could swing a scythe cut among the thickets weaving to the summit. At narrow switchbacks curves, shadows of claw-like branches of dead trees brushed my face and naked arms with ink-colored war paint. Suddenly, instinctively, I slammed foot on the brake, eyes affixed on a steed deer that stood frozen in a commanding stance on a nearby ridge, his head askance, his yellow-brown eyes leering into my gaze, the tips of his six-point crown antler piercing the indigo California sky. He took three proud strides, his beige coat ruffling in the wind, beckoning me with his eyes. A rare sighting of a deer in the wild, I felt compelled to open the door, race toward the hillside, tracking a spooked deer's ascent through the thick chaparral, soon meeting him by a ravine, where we'd listen to flitting birds and the sound of water slapping rocks in an invisible stream. But the deer gracefully turned his majestic head, projecting his yellow-gold ears like fur leaves, nodded, and pointed his antler toward a narrow path only he knew, leading me into a new world.

My trance was broken by a county station wagon making its descent suddenly swirling around a boulder in the road; saved by a firebreak carved off-road by a work crew. "Fuck me!" were my first spoken words since starting the eighty-mile journey from my Seal Beach home to L.A. County's Santa Monica Mountain top camp for wayward children.

I skidded the car to a stop in a gravel parking space in front of an elongated, one-story, concrete-block building painted desert-pink. Silencing the reeving engine and turning the key, I listened to riling leaves of nearby trees, pulleys clanging against flagpoles flying Old Glory taming California Grizzly Bear posted on each side of the double glass pane door entrance. A border of small rocks encircled the base of each pole. Next to a rock border, a varnished wooden sign bolted to two prominent posts boasted the burnt dark-engraved letters: ***CAMP VERNON J. KILPATRICK***.

Buffalo Dog

The moment I exited the vehicle, the ugliest dog I'd ever seen crawled out of its dirt bed under a manzanita plant by the front door. His dirty, matted fur showed white and light-brown blotches. His head and shoulders were large and square and disproportionate to the rest of his body. A tapered rear-end, male genitalia of a feral dog, and a short tail with a fur ball at the tip gave the appearance of a toy buffalo. Maybe it was. He didn't bark.

I followed "*Buffalo Dog*" through the double glass door of the administration building leading to the control center occupied by a stocky Black man chewing a toothpick. He was standing behind a counter turning a large sign-in ledger on top of a lazy Susan. A ball-point pen was attached by string to the spine of the ledger. He was dressed in a gray jersey marked, **Property of LONG BEACH CITY COLLEGE Athletic Dept**., faded Levi's, and GI-issued clodhoppers as I'd worn as a kid. He reached out his sausage-size fingers, and gripped my hand like a vise, releasing iron hold when I stiffened my arm and edged forward giving no quarter.

"Newbie," he announced in a joking tone to a slender white guy within earshot sitting at a desk in the adjacent office.

"Jerry Wheeler," I said, ignoring his tone.

"Raymond Dorset, but Dorset will do," he said, handing me the pen; his dark eyes hinting compassion for a neophyte about to face a baptism of fire.

I signed in the military, 0640, shone on a giant wall clock, on the page marked, Wednesday, October 12, 1966, and followed Dorset into a side office. He introduced me to an Italian-looking man, rising in a swivel chair behind a metal desk. He was dressed like a lumberjack in a plaid flannel shirt, *Lee* jeans, and high tops, but appeared underweight. "Sam Racetti," he said, extending his left hand awkwardly across his chest toward me to exchange a handshake. I noticed the last three fingers of his right hand were missing, giving his thumb, index finger, and palm the shape of an imaginary pistol. Averting my eyes from his deformity to two wall charts above the desk. The first was a table of organization (TO) starting with Director Virgil Kirkland, Assistant Directors Sam Racetti and Joe Becker, Administrative Assistant Shirley Rawlings, Nurse Joyce Saunders, 21 DPOs listed by surname, and 3 Recreation Leaders. A second chart, written in grease pencil, was a 62-hour shift DPO duty roster by Dorms: A, B, and C.

Racetti said, "Downtown did not give us your start date. Director Kirkland arrives at nine-ish. I'll go over the schedule after you meet with him. Meanwhile, Dorset will give you a brief tour. You're just in time for morning chow."

Standing outside the Administration Control Center offered a panoramic view of cinder block buildings bordering part man-made and naturally carved desert basin surrounded by mountainous terrain covered with boulders, rockslides, rain-starved trees, and thick chaparral with no visible trails other than a few narrow firebreaks. Dorset pointed out a baseball backstop and makeshift diamond directly in front of us and to its right, a football field of more sand than sod, a crude cedar running track whose northern side edged up to a grassy knoll leading to an elevated terrace with a posted sign: ***VISITORS AREA.*** This area included redwood picnic tables and benches shaded by ancient trees, constituting one-fifth of the campus interior property, The Visitors Area abutted a sagging 8-foot-high wire fence sans razor wire referred to as the "northern perimeter." Dorset gave me the heads-up why kids call Camp Kilpatrick, *"Krazy Kamp."* It was the only County camp with a boundary fence. He pointed at a rectangular building named, "C-Dorm," directly opposite the admin. It had a pyramid-shaped sheet-metal roof, an open double-glass door entrance, and several windows shattered by vandalism and neglect. He explained that C-Dorm housed 40 residents officially referred to as "students," ages 12 to 13. Next door, separated by weeds, was its architectural twin-B-Dorm- reserved for boys ages 14 to 15. A smaller concrete block building, "A-Dorm," on the admin side of campus, set directly across from B-Dorm separated by a football field size grass-covered ground. A-Dorm had secured key entrance metal doors and undamaged wire-mesh-covered glass windows. Dorset explained it was reserved for new admissions awaiting a vacancy in B or C-Dorm and used as a discipline dorm that had 5 solitary cells, 10 open private bedrooms, and the camp's infirmary. The tallest campus structure southwest of C-Dorm was the Gymnasium. The largest building was the Mess Hall referred to as the "southern perimeter" of campus property. Full-time teachers were hired and overseen by the local education district. The on-campus school classrooms adjoined the probation administration offices. All glass windows except B and C- Dorms were intact, and the grounds were well-policed of trash.

Anxious to meet the Director, DPOs, and students, I followed Dorset into the Mess Hall where we encountered a half dozen mostly white students wearing white paper garrison-shape KP caps, pressed white aprons over county work shirts, Navy dungarees, and county boots, bodies gyrating around wooden octagonal tables with four empty seats connected to metal base anchored to the shiny linoleum-covered concrete floor, their scrubbed hands retrieving silverware from dishwasher baskets, placing sets of knives, forks and spoons. I was impressed by the KPs' efficiency in setting up the mess hall for the morning onslaught of ninety hungry youth. Soon our attention turned to a pair of KP hanging like monkeys on a white muscular man's extended bulging biceps and arms, screaming, "*Popeye***,** Lift me! Lift me!" A demand the barrel-chested young man obliged with a grin, swinging his charges like rag dolls at a supply of tumbling napkins and condiments. The casually dressed medium-height man had short light brown hair, beady eyes, a boxer's face, and physique.

Dorset said Kilpatrick students assign staff a unique moniker inspired by the subject's personality, appearance, or mannerism. A moment later, sipping the first cup of Kilpatrick coffee and dipping toast in over-easy eggs, with an ex-Jr. college linesman, *"Bouncer"* Dorset, and ex-UCLA weightlifter, Greg *Popeye* Benny, I wasn't disappointed.

Child Army Of Meandering Misfits

Groups of young adjudicated delinquent boys hooting and hollering, started filing out of B and C Dorms, extending their arms to establish the proper marching interval. It was like watching slovenly recruits of child army misfits, stocking caps askew, pulled over eyes or wadded in hand, a few wearing oversize, unsnapped sand-color fireman's jackets, wrinkled denim shirts, baggy dungarees buckled above the waist, hanging loose, tattered cuffs dragging behind unlaced county boots. After repeated awkward attempts at precision steps to form a column of twos, the intervals collapsed like an accordion; marchers kicking each other's heels and playing catch with stolen stocking caps. A few columns, attempting corrective turns, exchanged love taps on the shoulder while others pushed the person in front with both hands and spun around reacting to the neighbor playing grab-ass or trying to stab him in the balls. Others, catching up with bunkmates, surreptitiously retrieved stones from their pockets and pelted sibling rivals for yesterday's insults. Meanwhile, KPs beside themselves gawked out the window at the child army of sad sacks racing helter-skelter across the field of St. Augustine grass and weeds buried in a loose layer of fine desert sand and patches of asphalt. A Black C-Dorm kid, raising his fist in a black power salute, was pulled from the line by a white DPO and then escorted arm-in-arm to the time-out bench. He joined a white knucklehead benched for throwing rocks at the group in front of him.

"Fuck me!" I mused.

After witnessing their extreme horseplay from my bird's eye view from inside the mess hall, more blind spots conducive to escape jumped out at me. The fence behind the gym was out of the staff's line of sight. Notwithstanding, feeling like an interloper observing mayhem, seeing DPOs' "all hands on deck" strategy turning pandemonium into order: rounding up strays and stragglers, and herding wild colts like experienced rovers into parade formation, without threats of violence or physical force, gave me pause. Meanwhile, a kid carrying a folded flag exited Admin with Assistant Director Rachetti. They stopped at the flag stand. The student attached the America's flag to a rope suspended on a pole, then slowly raised Old Glory to the cacophony of clanging pulleys, shuffling boots, and incoherent voices of two platoons of child prisoners of society's war against children reciting the Pledge of Allegiance to stars and stripes flapping for attention in the desert sky.

Fifteen minutes later, boys started regrouping in front of the mess hall to march back to dorms and prepare for inspection before the next movement to school across no-man's land.

Dorset led *Buffalo Dog* and newbie Senior DPO down an asphalt walkway toward A-Dorm. A kid from the barrio sweeping waves of sand out of the path with a push broom paused and stepped aside. The dorm sat on the camp's northeast section of the campus property overlooking a football field and Visitors Area. A young, blond, white DPO with a ponytail, wearing a bead necklace, psychedelic t-shirt, stone-color jeans, and custom-made leather sandals, stood wearily at the base of a nearby tree, coaxing a kid down. The climber, a short, dark skin Hispanic kid in a stocking cap, County fire jacket, clothes, and boots, stood on a limb ten feet off the ground, one hand grasping a branch above his head, the other, the zipper of his fly. He warned, "Any closer, hippie, I'll pee in your face!"

Ignoring the tree climber, Dorset unlocked A-Dorm's door, the mute *Buffalo Dog* scampering inside ahead of us, past a door to the infirmary, flying over three steps, landed on his belly in the middle of the dorm's elevated control center, then paddled across the slick floor, legs spread-eagled landing at the feet of a tall, white man standing by a desk, facing the dayroom of young hellions below.

"You must be the new DPO," said the tall, tanned face man, eyeballing me with his shifty steel-blue eyes. He appeared my age, had a burr cut, paratrooper's shoulders, and wore a smudgy T-shirt, straight-leg jeans, a wide leather belt, and high-top sneakers.

"Jerry Wheeler," I confirmed, exchanging strong handshakes.

He said, "KC Barton, I am about to start my orientation spiel. You're welcome to listen."

The U-shape control center was partially enclosed by a Formica countertop overlooking a spacious dayroom that included three octagonal tables anchored on a concrete floor occupied by five "newbies:" three white, one Native America, and one Black. Standing by Dorset, I noted the control center afforded the dorm supervisor an excellent field of vision. Across the dayroom, an elongated tin urinal, two semi-private commodes, dual washbasins, a wall mirror, and three open showers came into view. And seeing the entrance of 10 semi-private bedrooms in the southern half of the dorm caused no strain. All the walls were painted puke green.

KC scanned the group of restless child prisoners trapped on desert mountain Devil's Island. One kid's legs were so short the boots of his straddling body did not touch the floor.

KC began: "Morning, Gentlemen, my name's KC Barton. Most of you know Mister Dorset. The other gentleman is DPO Wheeler. He has something in common with you. He's the new guy on campus, too. We know you have heard negative rumors at Juvee that Camp Kilpatrick is a terrible place because it has a boundary fence."

"Yeah, *Krazy Kamp*," a kid interrupted.

KC continued, "Given that the State of California and County of Los Angeles are morally and legally responsible for the safety and welfare of children and young minors, the authorities deemed it appropriate to erect a boundary fence on the summit of a desert mountain to keep hungry mountain lions out. The second rationale for the fence is to give anyone contemplating AWOL to reflect on the risks of encountering nature's hazards going down the mountain alone: broken legs falling sheer cliffs, cuts by jagged rocks, fighting one's way through thickets with thorns sharp as needles, becoming lost following deer trails leading to dead-ends, blind canyons and dark caves full of blood-sucking vampire bats, rocks hiding venomous rattlesnakes or being stalked by a hungry mountain lion ready to leap on its next meal. Because of these dangers, authorities only approve search and rescue operations in daylight after you are declared missing for 24 hours. Further consideration, if you get itchy feet, scale the fence, and don't voluntarily return within a reasonable time, say climb back in or knock at the front door within two hours, it will cost you two days in lock up, a job demotion, and two-week camp restriction."

After a long pause, reflecting on the boundary fence's nature's warning and consequences of failure to return, KC continued in a business-like tone, "This is A-Dorm, the dorm for newbies that offers room service and has its field period. You'll be here until space is available in other dorms. A-Dorm is also called the adjustment dorm because it is always locked. You'll meet your own DPO counselor when transferred. Mr. Dorset and I are Recreation Supervisors that do everything a DPO does except have a caseload because we haven't finished college. Every student including newbies attends school and has a paid job assignment. If you make it through a day without getting a misconduct report, the county will credit your camp account sixty-five cents. However, if you engage in disorderly behavior like throwing stones, fighting, race talking, going out of bounds, kicked out of class, you'll spend the day in A-Dorm, miss work, and docked pay. And risk being demoted to a lower job. The work assignment board is posted in the front office. Each student is permitted one set of personal clothes to wear off-campus. How do you get off-campus? First, your counselor may take you to the beach or field trip. Second, one day a month, their family can take them off the grounds. Third, teachers and DPOs evaluate students' weekly academic and program progress. The top-ranked eighteen students at the end of each month earn a ticket to a first-run movie in Santa Monica Friday night on the camp bus. The fourth way out of here is to be transferred to YA as a screw-up. Tomorrow, the Camp's Student Mayor will walk *newbies* through the program and explain the details. Finally, brand the five commandments of *krazy kamp* in your brain: **(1) "No race talk; (2)No mother talk; (3)No throwing rocks with lethal intent; (4)Big guy don't fight small guy;(5)Violation of off-campus honor code is expulsion."** *Buffalo Dog*, lying on his side next to KC's feet, slowly aroused. The dog looked upward at the chubby redheaded kid wearing only *fruit of looms*, surreptitiously listening to KC's orientation spiel. Suddenly a kid stepped out of the entrance to the lockup section, smiling, winking, giving us the finger, then casually slid into an empty seat in the dayroom with newbies. The next voice we heard was at the entrance of A-Dorm behind the control center.

"Keep your fucking hands off me, *Tippy Toes."*

It was the same tree climber that threatened to pee on a DPO hippie. He was now escorted by him and a skinny white guy in his early 20's dressed in a sweatsuit and low-cut sneakers the kid called, *Tippy Toes*. Each captor held the pint-size culprit a foot off the ground by an arm, as far away as possible to avoid being kicked in the side or legs by heavy swinging clodhoppers. The resisting child prisoner wore a navy-blue stocking cap pulled over his ears and an oversize county-issued tan color canvas fire jacket with large unhooked black-metal snaps. His tattered-cuff jeans fell over his boot heels. He appeared under 5 feet, and too young to be a teen.

Huffing and puffing like a blowfish, carried inside the dorm, he spat, "You fuckers ain't sticking me in lockup again. I wanna see my counselor."

"William's off duty, Travis," Dorset said, as Travis wiggled out of DPO *Hippie*'s grip, whirled around, ripping off *Tippy Toes's* silver wristwatch, stretching the retractable band into figure 8.

"Easy, Travis," *Tippy Toes* said in a pleading tone. "That watch's ain't **County**."

"Tough shit," the kid replied, hurling the watch at A-Dorm's concrete wall, smashing the crystal to smithereens.

"Damn it, Travis, that watch was personal," *Tippy Toes* said, tensing his body, "Gonna cost your butt a week in lockup."

KC's brood of newbies', and *fruit of the loom* kid's eyes alternated from KC's stern face to the kid dodging Dorset's clutching hands.

"Why are you guys picking on me? I didn't do nuttin'," Travis protested, now caught in Dorset's bear-hug.

"Nothing but peeing your bunkmate's boots and causing a riot in the dorm," said *Tippy Toes.*

"Well, shit, he was mother-talking me." "Zip it, Travis," KC ordered, motioning Dorset to drag 60 pounds of trouble past the control center to a seat below. "You know the procedure, Travis. Start spittin' again, gonna lock the cell door until William's on duty."

"Next goddam week," said *Tippy Toes* in a vindictive tone.

As Dorset released Travis, the kid shook his head wildly, then mimicked *Tippy Toes's* sad face grieving over his smashed watch.

"Okay, okay, KC, I can undress myself," Travis insisted, unsnapping the fire jacket and throwing it on the floor, unlacing and removing his boots, throwing them blindly at *Hippie* and *Tippy Toes*. Travis looked at me and *newbie* rubberneckers watching his antics, and smirked, "Welcome to *Krazy Kamp*, motherfuckers." Then proceeded to pull off and discard his jeans in a heap on the floor, ripping off his denim work shirt, popping buttons one at a time, throwing it over the control center, followed by removing and tossing his stocking cap over his shoulder as he strolled to a lockup cell in the backroom, talking over his shoulder, "KC, gonna lock me in or give me some slack?"

"What do you think?" KC said, retrieving an oversized brass key tied to a wooden handle by a lanyard, protruding from his hip pocket.

While KC secured his newest guest in lockup, and Dorset watched *newbies* and a half-naked intruder nonchalantly sitting with his arms crossed, I mulled over the effective DPOs' double-team physical force technique getting Travis to A-Dorm. I also understood why the captor wearing sneakers was called "*Tippy Toes*" when he bounced on the tip of his sneakers like a ballet dancer exiting the dorm, swearing over a watch whose time had run out.

Curious, I followed KC to lock up and observed him slow-walk Travis into an 8 by10 foot room lit by a wire-covered ceiling lightbulb and natural light from a single wire-mesh-covered glass window facing the mountain landscape. The cell contained a clean bare mattress on the floor and a folded Army blanket. Travis scrunched his face as if he'd been there before. It was the first time I'd seen a kid in solitary confinement since *Boys Industrial School* in Lancaster, Ohio.

KC said, "Travis do you need a head call or a drink of water?" He was met with silence; KC closed the heavy iron door that had no peek hole, inserted it, and turned a large brass key in the lock that sounded like closing the cell door of a medieval dungeon.

Reading my raised brows, KC said, "Don't worry, Wheeler, he knows I'll let him out soon in the dayroom."

"So, what's the lockup cell policy?" I asked.

"Unless a kid's violently aggressive, it's up to the staff that wrote him there or on A-Dorm duty. I tried to keep the cell door open and used it as a timeout room, like the redhead sitting out there in his underwear. He started a food fight at morning chow. Lock cells are checked every 15 minutes."

Broken Windows With A View

Wesley Green, the camp's boy mayor, an honor bestowed by staff to the most outstanding student on the discharge list, led me and five newbies across the main recreation area to an empty dorm during noon chow. "Fuck you, Green, you big kiss ass," said the B-Dorm rock thrower being escorted to A-Dorm. The boy Mayor responded by pressing his right hand across his upper chest, and saying, "Heartache, your momma sucks dog bones."

B-Dorm was lit by sunlight coming through cracked or missing windows wrapped around a spacious concrete block building divided into two open areas with 10 bunk beds and space for showers, wash sinks, wall mirrors, semi-private commodes, and metal trough urinal. Each bunk bed was covered with an Army blanket showing a white sheet collar and puffed pillow and an extra folded blanket at the foot. Two brown-painted metal footlockers sat at the foot of each bed. Some displayed wooden and plastic model planes, cars, and books. Standing in the center of the elevated control center afforded a 360-degree vision of the dorm's interior which revealed adequate space for ping pong tables and dayroom amenities. The camp boy mayor described the organizational structure of the dorm.

"Each dorm has 40 students, broken down into four groups. Each group is assigned a DPO. Each group elects a boy leader. Some of you will have the same DPO as your individual and group counselor. Your DPO counselor approves all your program assignments and recommends your discharge. If you screw up, best to do it during his shift. Any staff here can put you on restriction in A-Dorm, but only your counselor can negotiate to get you off. So, mess up when he's scheduled off, you're fucked until he's on duty unless you get some slack from the DPO who wrote you up."

"What if your counselor doesn't like you?" asked shy, 14-year-old newbie, Max Yeager, standing in the control center, favoring his gimpy leg. Blond hair halfway over his ears and skin the color of butter, KC gave him some slack at shower call letting Max shower last, alone because Max did not want others to see his grotesque leg scar.

"You're fucked," Green said.

"What if you don't like him?" a kid asked.

"You're fucked," Green said. Adding, "Watch out for DPO Wildman Morris, He's got flaming-red hair and a hot temper. Orders his C-Dorm boy group to run laps around the track of the recreational field if they lose at volleyball. But, if you're in his caseload, Morris will take you on a treat to the beach."

"What this honor code crap?" a voice asked.

"Going AWOL on an off-campus visit with parents, a DPO, or the Movie bus. Because the off-campus dress code is personal clothes."

Butterfly Man

Mrs. Shirley Rawlings, Director Virgil Kirkland's matronly white secretary opened Director Kirkland's office door, and said, "Director Kirkland will see you now."

Kirkland was hanging a pale blue suit jacket on a vintage coat stand in the corner opposite the entrance next to a long table displaying a pair of glass-encased butterfly collections. Above the table were framed colorful butterfly prints. His dark eyes glinted with suspicion and relief, reflecting a hint of resentment toward his boss Sam Ostroff for denying him the courtesy of participating in the hiring decision of Senior DPO. Shaking hands, we stood eye-to-eye, separated by two decades of perspective on the juvenile justice system taming rebellious youth. Kirkland appeared this side of middle age handsome, clean-shaven, shoulders losing the battle over hunching over government manuals, books on theories of changing adolescent behavior, news on rising counterculture, psychedelic drug revolution, anti-Vietnam war protests, and saving the endangered Monarch and Painted Lady Butterflies.

Looking out a one-way tinted glass window offering a panoramic view of the camp's interior showed DPO Benny escorting KPs to the Mess Hall, other DPOs standing by, getting psyched for battle stations monitoring mass noon chow movement, Kirkland said, "Sam, must've been impressed by your work, and we're sure ready for some new ideas," Didn't see Kirkland again until the bi-weekly staff meeting.

Fourth shift. Standing between Camp's Director Kirkland and Sam Ostroff, *Godfather* of L.A. County's Juvenile institutions, I was unlocking A-Dorm's only occupied solitary cell. The Godfather was holding a freshly groomed, shivering miniature snow-white French poodle so scared of *Buffalo Dog,* KC locked the camp's mangy mascot in the A D's office. This was Ostroff's first "surprise camp visit" since he swore me in.

"What we have here, Sam," the *Butterfly* man said, our eyes focusing on short, wiry half-naked, shaggy brown hair, tan face 14-year-old B-Dorm escapee staring through a wire-mesh window, "is a young man giving us the silent treatment."

"How long has he been keeping this up?" said Godfather.

Kirkland replied, "The lad hasn't uttered a word since KC tracked him down hiding in the brush outside the boundary fence behind the gym. Sulking over Morris, his counselor kicked him off their volleyball team for "losing a game." Refuses to talk to Morris or anyone here."

"Don't you think it about time you talk to someone, son?" said the red-nosed mobster, gently stroking the poodle.

"Have anything to say?" Kirkland asked, trying to be kind. Holding an Army blanket across his shoulders like a cape, Randall Cunningham, a child prisoner, slowly turned, directed his dark eyes at

Camp Kilpatrick's Commandant, and said in a raised voice, "Fuck you!" Shifting his gaze to me, added, "Fuck, *Pushbroom* here." Then shouted at the *Godfather*, "Fuck your friend," which ended the interrogation, eyes locked on the shivering poodle cradled in *Godfather's* arms. "And Fuck your friend's dog," covering his head and shoulders with a blanket cape, executing 180-degree-turn, staring again out the window at scurrying birds in mountain brush.

After a moment of dead silence, a raspy voice with cigar breath said, "He's ready."

I immediately observed this was a mismatched student/counselor assignment. Morris was depicted by his DPO peers and students as a quick-tempered *wild man* and a poor loser. This warranted transferring Randall to my caseload and supported my proposal of transforming A-Dorm from a primary disciplinary unit to an Intake and Orientation program for assessing *newbies* and matching them with appropriate counselors.

Consistent with A.S. Neil's belief and philosophy that children understand and respect adult boundaries if authority figures are candid in educating the risks and consequences of violation. Henceforth, in my orientation, newbies were informed that a special camp to help children in trouble established on top of a mountain surrounded by a rugged landscape of rocks and canyons and dangerous lions will be closed under public pressure if one of its kids was seriously hurt running away. Search & Rescue takes time and nature is not always kind. Running away during a field trip jeopardizes future off-campus activities. DPOs are not required to take students off-campus. It is not in their job description. More importantly, students had to weigh the personal alternative consequences if the county closed Camp Kilpatrick: Juvenile Hall or transfer to Youth Authority, institutions with walls and no student honor system.

Baby Giant And A Pandemonium

I unlocked the dorm facing *Tippy Toes* and a pair of eager beavers: a *newbie*, and a kid on restriction. The former was a dark brown, curly hair, bright-eyed, white "tall glass of water," John Blake, a month shy of age 15, from Long Beach. According to his case file, he had a history of disruptive behavior in school, e.g., throwing textbooks out the window, and spending days at the principal's office. No record of physical violence saved him from YA. The Juvee shrink noted that Blake had high test scores and extremely low frustration tolerance. I found the shrink's 10-minute standard diagnosis, "Reaction Formation to Adolescence," a useless esoteric term attributing nonconforming behavior to the age-specific teenage syndrome. The minute newbies took in Blake's gangly, six-two measure when he entered the dayroom, emptying a shoebox of science fiction paperbacks from home including Robert A. Heinlein's *Stranger in a Strange Land*, a voice gasped, "Jesus Christ, it's ***Baby Giant***!"

The other adjudicated youth, Douglas Henderson, had a proclivity for smoking dope, dropping acid, and "communicating with the dead at a seance in school." He attempted AWOL from C-Dorm on his 15th birthday, the same day his mother, a medium and palm reader in Hollywood missed a scheduled camp visit. Teachers thought Doug was crazy, even though his mom made a good living at it. He could recite anyone's horoscope from memory. His first request was to get back his tarot cards "*Wildman*" confiscated after Doug put a hex on him for catching him outside the boundary fence where he stayed several hours until recreation leader KC Barton, talked him back inside. His second request was a new DPO counselor which I obliged.

The next morning, I awoke in a dream seeing *Vincent van Gogh's images of children shuffling in circles in a prison yard.*

Visitors Day arrived to my chagrin. Counselors wear boots, and old jeans, put their glasses and watches away, shrug their tired shoulders, and arm themselves with bandages and Mercurochrome to battle the rejection and rage of boys without visitors. It was also the day boys scurried to dorm roofs and trees to hurl stones and expletives at skedaddling or incarcerated dads, and moms. Those who had no visitors because their families were too poor or busy for the long bus ride to the mountain top with younger siblings desperately wanted more love than their parents could give. Outcasts climbed to the highest oak, scanned the unfair world that they did not create, and pissed on the blessed and themselves. A few escaped self-hate, scaling wire until caught or talked down by staff who understood their pain. Even the meekest prisoner who experienced no visitors, agitated peers to blows, spat, backtalked, and lashed out at staff until dragged kicking and yelling to lock up out of earshot of a mother's soft voice. Getting angry teens out of sight of large picnic baskets overflowing with guilt and kindness, or a father's strong arm became routine. DPOs learned to mitigate rebellion by targeting *Krazy Kamp's* "visiting day orphans" contemplating scaling or clawing to freedom surviving on streets without love by keeping them out of sight of the *Visitors Area* during visiting hours, organizing basketball games in the gym, artwork, card games, model building in dorms, and if possible, nature hikes and field trips in Santa Monica.

The unanticipated threat to adolescent struggle for self-identity and to social order in coercive remote mountaintop setting on otherwise a normal Saturday occurred while students were lining up in dorm groups for noon chow movement. Suddenly boys fled their formations, scattering for trees, rooftops, and the boundary fence. We had never seen anything like it. A fucking mass mutiny under my watch as Senior Deputy Probation Officer-in-charge of *Krazy Kamp*.

"What the hell is going on, Cowboy?" I yelled, watching him run back inside B-Dorm to retrieve his cowboy hat, then race to the boundary fence.

"***Shears***!" Cowboy shouted, straddling the fence, waving his hat, kicking his boots, as eighty of the county's zaniest misfits broke formation, shouting, "***Shears*** is here! Run!"

"Shears?" I asked.

"The county barber," Cowboy laughed. "Only knows two kinds of haircuts--fuckin' buzz and bowl. Nearly scalped and cut off a kid's ear last time."

"What happens now?" I asked myself, observing a nearby kid, grimacing, running his fingers through his precious brown locks. I spotted a few sheep further down acting like they were being dragged to slaughter.

"Looks like they caught the poor KP's and the Broom Sweepers," a child's voice said sympathetically, hiding behind a bush next to B-Dorm.

"Mother Fucker! Mother Fucker!" chanted a dozen wards perched in trees and on top of baseball backstop.

"Don't worry," a veteran C-Dorm boy group leader said. "Shears only stays an hour 'cuz he cuts hair at other forestry camps. "We'll just wait him out."

He was right. Six unlucky shorn sheep wearing stocking caps to cover their shame returned to volleyball, shooting hoops, playing catch, and cards, strumming guitars with broken strings, reading paperbacks, and sprawled in the grass, daydreaming in suspended time.

By mid-November, Max was coming out of his shell. He worked his way up from sweeper in A-Dorm to KP in B-Dorm and was less self-conscious exposing his leg disfigurement, resorting to "speed showers" and earned one trip on the movie bus. His parents made bi-weekly campus visits and counseling sessions. My goal was to recommend discharge by Christmas. All things considered, Max was fortunate to have supportive parents that wisely rejected inpatient psychiatric options appropriate for treating psychotic disorders. Both were devout Christian Scientists, rejecting placing him in a psychiatric hospital because drug therapy was against their religion. Consequently, Max's "crime" of truancy led him to Maclaren Hall.

Max was also lucky. A fellow MSW, Jack Taub, Cigar Face's point man at Maclaren Hall, screened and selected candidates for *Krazy Kamp*. Jack used his clinical experience counseling street kids experimenting with drugs, school dropouts, outliners, and runaways. Max's world crashed the day his right leg was mutilated, horse playing on an oil well pump jack, requiring multiple surgeries and months in physical rehab to barely walk. His life of competitive swimming, hiking, playing sports with peers, and making honor roll grades, were obliterated. He became a home recluse, refusing to return to school, psychological counseling, or participate in homeschooling. According to his parents, he only wore pajamas and stayed in his room. His father was a federal civil servant, and his mother was a stay-at-home mom. Taub attributed Max's social withdrawal as a traumatic reaction to his injury inextricably linked to his self-image exacerbated by adolescence. Clinically, Max appeared an atypical candidate to heal physically and psychologically in a remote mountain setting established to tame a rambunctious youth rebelling against societal norms. But Jack's final test of sleight of hand magic card tricks broke Max's silence. Jack related the extraordinary heroic story of legless Royal Air Force flying ace Captain Sir Douglas credited with 22 aerial victories during the Second World War.

DPOs and Recreation Leaders were exhausted by the shifts' end. For me, the shift's end started Monday, at 7 p.m. with a jolt of caffeine, a two-hour-plus ride home, crashing on the couch with a cold beer, and 90 lb. female German Shepherd, Braun, pawing me for attention while letting my wife sleep undisturbed before reporting to her teaching job at Long Beach Lindbergh Jr. High in an impoverished community.

Camp Kilpatrick was ruled by custom and tradition. Staff turnover was nil. No policy manual governed dorm protocols and campus activities. Education curricula and faculty were accountable to the local education district. Traditionally, A-Dorm was considered a short-term, secured way station for new arrivals and a "time out" dorm for general population students cited for misbehavior. Each DPO was responsible for completing a progress report and discharge recommendation to the court on students assigned to his caseload by one of two Assistant Directors: a policy referred to as matching individual youth to appropriate DPO "counselor." A function delegated to yours truly Senior DPO overseeing the newly started *Intake & Orientation Program*. The average DPO counseling caseload was five students. The average length of stay was six months. Each B & C Dorm resident was assigned to a DPO associated with the same dorm. Individual DPOs were responsible for his caseload's family counseling sessions which rarely materialized due to geographic distance. DPO planned campus recreation activity for one of his dorm's four, 10-member boy groups. However, in any given shift, all DPOs on duty made up a "backup" pool to fill in other dorms including A-Dorm in an emergency. But "backup" duty accounted for only a fraction of one's shift.

I was leaving the mess hall after morning chow when KP duty supervisor, DPO Joe Brown accused KP, Miguel Yazzie, a 15-year-old joyrider and runaway, two weeks from discharge, of stealing utensils from the mess hall. If written up for "possession of contraband," Yazzie would be demoted to "Sweeper," and jeopardize his graduation date. Students called DPO Brown, "*Crater Face*" behind his back for trying to conceal crater-size zits on his face with a scraggly beard, and because he had a reputation for cuffing "back-talkers" to the punishment bench for hours until the practice was banned at my urging. He also favored using lockup for out-of-bounds violators.

"Mister Brown, Yazzie's got contraband," said the KP accuser, jealous of Yazzie's job board status, frequent movie night award outings, and popularity among peers. Yazzie, a Native American born in the capital of Navajo Nation, Window Arizona, and grandson of WWII Navajo WindTalker, was raised by a mix-Anglo Navajo mother and Hispanic father that migrated to Los Angeles for work.

"What's going on, DPO Brown asked the accuser, from behind dark sunglasses, a LA *Dodgers* baseball cap hiding his forehead.

"I'm short of a knife and fork and I don't want to lose my job, Mr. Brown.""I didn't steal nothin,'" Yazzie protested.

"Let me search 'em, Mister Brown. Betcha I'll find them."

"Okay, Yazzie, empty your pockets," ordered Brown.

"Okay, okay. I don't have shit," Yazzie said, turning his fire jacket's pockets inside out as a pair of silver metal objects conspicuously flew out and skidded across the shiny beige linoleum floor.

"See..., See.., What's I tell ya?" said the KP.

"Fuck you. You planted them," Yazzie said, pushing the accuser off balance.

Before the KP recovered, Brown stepped between them, and grasped Yazzie by the collar, shoving him out the door.

"On the bench, **Chief**," said Brown using Yazzie's fellow students' non-pejorative moniker, but in the tone of a racist slur. Yazzie was often called upon to lead hikes outside the boundary fence because peers and staff alike were confident Yazzie's tracking skills would get them back home. "I didn't do it," Yazzie said, holding his ground, looking at me to support his claim of innocence. KPs stored their fire jackets on wall hooks, during work; a convenient time to plant evidence.

"Move it," insisted Brown, pushing Yazzie toward the punishment bench. Each time the DPO pushed him using more force, Yazzie resisted harder. Brown grabbed Yazzie by the back of his belt and dragged him back to the punishment bench.

Yazzie responded, clawing at his captor's hands, and kicking him in the shins with his clodhoppers, prompting Brown who outweighed the boy by 30lbs, holding him by the belt, whirling him with one hand, and smashing Yazzie across the mouth with his forearm.

KC, passing by, looked without expression at Yazzie's bleeding mouth, and paused, handing him a hanky.

"Don't do me any favors," Yazzie said.

"I won't," KC said, walking away, placing his right hand across his chest, giving Yazzie the "heartache" sign. Meanwhile, *Buffalo Dog* paused, lifted its leg, and peed on the bench, spraying Yazzie's boots then ran to catch up with his master. While waiting for the infirmary to open for sick calls, we had a ringside view of a post-noon chow movement to the dorms. All available personnel exited admin and fanned out in the recreation area. During the next five minutes, at least a fifth of the students got pulled out of formation for engaging in horseplay, rock-throwing, and agitating staff, and peers. Three boys bolted for the trees on *Visitor's Hill*. Another kid, Jeff Kemper, a.k.a. *Cowboy*, scaled to the top of the boundary fence, screaming profanities, and waving a cowboy hat. Another kid mysteriously raced across the recreation field from his B-Dorm group to A-Dorm, beating on its locked entrance, demanding to see KC overseeing the dorm. Others that fell out as stragglers, take a shortcut across the field to get to their dorm before a DPO carried out the threat to write them up.

Tippy Toes casually strolled to the base of the boundary fence, entreating *Cowboy* to come down. A task hindered by tree climbers shouting, "Run, *Cowboy,* Run," as they swung from branch to branch like wild monkeys.

Reacting to the ruckus, Tom Gilder, alternate Assistant Director filling in for Racetti, exited admin, brandishing a yellow write-up pad. He spoke with a Dutch accent, had dirty blond hair that grew in every direction, and a big square head. Students called him, "*Frankenstein*," mimicking Boris Karloff's stiff walk to get his goat. Probably hoping that none of the trio had the impulse to spit or pee on him, he wisely kept his social distance.

Meanwhile, Donnie Berryhill, a.k.a. *Choo Choo* the only Black DPO, that always wore an official railroad cap and overalls, used Admin's public-address system to coax the kid pounding Intake (A-Dorm's) door to join us on the timeout bench until KC was released from dorm coverage.

Exasperated by the standoff at Visitors Hill, *Tippy Toes* and *Frankenstein* sat on a picnic bench, waiting for reserves that appeared in DPO Phil Henry, a.k.a. *Hippie*, minus his bead necklace and headband, exiting admin. As *Hippie* walked toward the mini-revolt, he crossed paths with the Intake Dorm's intruder on his way to the "time-out" bench. The kid's name was Barry Grant a.k.a. *Tiny*, an immature 15-year-old red-head squirt from C-Dorm.

"So, what's the big deal?" I asked.

"Only model glue we're allowed to use in the dorm is fuckin' *Elmer's* glue, man. Same as using your damn snot. Only real glue for models is *Testor's* glue in the office, supervised."

"So, what's the hurry?"

"KC **promised** me yesterday. But he didn't let me use it."

When Yazzie and I reached Intake Dorm's entrance, we observed DPO *Hippie* coaxing the last kid down from a tree at Visitors Hill, and weekend Assistant Director (A D) *Frankenstei*n trailing two ex-Tarzans shilly-shallying toward B-Dorm.

Sick Call

Inside the Intake Dorm, Yazzie joined several malingerers queuing for a glimpse of *Krazy Kamp's* gorgeous young nurse, Joyce Saunders a.k.a., *Medic,* and a little dose of motherly attention. The next child patient came in with a crudely engraved infected ballpoint pen tattoo of lightning bolts, his pus blending with blue ink, forming green Zs on the back of his hand. This was the first opportunity to talk to our nurse about the staff's use of physical force and campus safety issues.

Ignoring awaiting patients' jealous hoots and howls, *Medic* Saunders let Yazzie jump the queue, probably because I was with him, and he was the only kid with blood on his face. The examining room had life-size wall posters of the male and female anatomy: pink muscles, purple arteries that look like tubes, red blood vessels, green veins thin as a thread connecting vital organs, white bones with Latin names, an examining table, a small refrigerator, locked medicine cabinet and a shelf containing medical supplies. Yazzie sat quietly on edge of the examining table, staring unabashedly at the young medic's slim hips, perfect round ass in tight-fitting jeans ending at burnt ropers, as she slipped her long hands into latex gloves, fetched alcohol and cotton from a medicine cabinet, turned hard body, advancing toward him. She had a tanned complexion, short, bleached-blonde hair, swimmer's shoulders, and blue-green eyes as bright as Egyptian amulets. She wore a pale-blue long sleeve denim western shirt with oyster buttons that accentuated the presence of natural bosoms in each breath. In my childhood, nurses didn't look like she did. Instead, they were old and overweight women in white uniforms and white hose that tended me at Doctor's Hospital six years ago. I was grateful I had an attractive, intelligent, half Swede, half-German wife waiting at home for me.

"Going to sting, Hon," the *medic* said, bending down, gently wiping dried blood off the young stoic Navajo lips and chin. Catching his eyes watching her jettison rubber gloves and jotting notes in the medical chart, she smiled and gave him the belated heads-up on *unwritten rules* on surviving *Krazy Kamp. "*First, always give DPOs extra slack during the last day of their shift. They are suffering **shift fatigue**, judgment impairment from sleep deprivation, withdrawal from sex, booze, and freedom. And, for God's sake, don't agitate the teachers. Only a few good ones left. And most importantly, talk problems out with your counselor. Most DPOs are good guys, and they're your ticket home."

Privately, the camp *medic* jokingly described the triage system for treating casualties of *Krazy Kamp e.g..,* fistfights, infected tattoos, and sprained ankles. The prescription for bedwetters was unorthodox but worked. At the big dorm, the bedwetter would tell his DPO that will apprise the DPO covering A-Dorm– the place where mattresses were stored– to expect a boy dragging a soiled mattress and sheets across a hundred-yard field to lock up is on his way. Upon arrival, the bedwetter will change shorts and sleep on a clean mattress in an unlocked cell until he returns the next morning to his dorm carrying a clean mattress and sheets, without ridicule or prejudice. The cure was usually a few trips.

However, she let me know, if illness or injury requires intricate stitches, x-rays, lab tests, emergency surgery or appeared life-threatening, the boy was transported to Santa Monica Hospital. Her greatest fear, and I agreed, was a boy injured or lost in a mountain brush attempting AWOL. She supported my newbie orientation, stipulating the only safe escape route was following the switchback road ending at Pacific Coast Highway, emphasizing nature's hazardous mountainous landscape in the pitch-black night darkness. She couldn't recall a kid ever making it. AWOLs wander back to camp, scratched up, bruised, and hungry within a few hours. The incentive to voluntarily return before we report a student "missing" to authorities and family is high. The maximum penalty for an AWOL's voluntary return was job demotion and a month's camp restriction. A newbie was warned upon admission that violating the off-campus "honor code" which always calls for escaping such events, events such as movie night or a group treat to the beach, constitutes a capital offense and mandatory camp expulsion. That is because staff unequivocally reject the principle of "group" punishment. Nurse Saunders speculated that DPO Brown was an outlier whose days were numbered after the Camp Director read her report on Yazzie which is required for a suspicious student injury. The medic's *Krazy Kamp* tenure proved invaluable in understanding the unique institutional cultural factors that mitigate against staff abuse. We all knew abuse was prevalent in traditional juvenile correction facilities both from the personal research and direct observation.

Yazzie's discharge date was never in jeopardy. Notwithstanding Brown's formal report alleging a model student's "possession of contraband," the crime made no sense unless a knife was used as a weapon of which there was no historical evidence. Moreover, guilt and level of punishment had to be concurred by Yazzie's DPO counselor Benny, reviewed at a general staff meeting, and approved by Director Kirkland.

Brown was administratively transferred to probation's field division. shortly after the incident. He got what was coming to him.

Voices

The personal typewriter, toolbox, and Sony Tape recorder do not have a county property number. A box sat on the table containing watercolors, half-pint cans of oil paint, brushes, various-sized stencils, tape, and scissors. Several poster boards and pieces of plywood sat next to an opened metal toolbox with an electric drill in the top drawer. Since the day was Saturday and no school was held, boys were headed to their rooms to take a siesta until going to the field outside.

I descended from the control center and was sitting on top of one of the day-room tables, using the seat as a footstool. I said, "We got to prepare this place for new arrivals; you're not spending the afternoon lounging in your room."

Four new boys stood, frowning. Melvin Hooks, a short, skinny black 13-year-old kid from Watts sat on a stool, nervously shaking his leg. A DPO told me Blacks started coming to *Krazy Kamp* after the Watts riot. Sitting on the floor with their legs crossed, facing me was David Green, 14, from Long Beach, dark-red hair always slicked back as if heading to church, and a short 15-year-old with pale skin, curly black hair-Herby Johnson-who spent his life indoors drawing weird pictures with colored pencils. The fourth new arrival, a muddy-blond crew-cut boy named Henry Ludwig who spoke with a German accent, was always the first to volunteer and had a fetish for clean clothes.

"Is this one of the group meetings you talked about?" asked Johnson, his thick lips slack.I stared at Johnson, then said, "Well, yes and no. If you want to hibernate in your room, Mister Johnson, go ahead. But I would like your participation in improving the physical appearance of our home here. I would like your ideas and to see if we can organize a little division of labor based on individual interests and talent. Like, I've seen Green here, drawings. We could sure use his assistance in making up some signs for the dorm."

"What kinda signs?" Green asked, doodling on a scratchpad.

"Posted activity schedule and dorm rules. Also brought some wall paint for your rooms."

"What you mean, rules?" Grunt asked.

"Rules for using a ping-pong table and a carrom board, to discourage bogarting and fights and breaking equipment, stuff like that. And a bulletin board to post daily activities. You like to play ping pong, don't you?"

Grunt answered with one short guttural sound.

Green laughed, scratching a sideburn. His blue eyes scanned the dorm. "Mister Wheeler, they not allowing that stuff in here. Besides, where we gonna get a ping-pong table and carrom board?"

"Well, I got a carrom board in my car. The ping-pong table will take some doing. I understand Camp Kilpatrick has a woodshop, so I guess you boys will just have to make one. And you do not have to stop there. We could use some bookshelves and tables for your rooms."

The group exchanged skeptical glances. "So, what happens to our stuff when we leave?" Johnson asked"You can donate it to the Intake Dorm, take it with you, or send it home. But before we go on, there is something I want to show you." I went to the control center and returned with a plastic machine the size of a portable record player with two reels of tape on top and several dials. I handed the cord to Hooks, gesturing for him to plug it in.

"This here is a genuine *Sony* tape recorder I want to use to record our group meetings. After meetings, we can playback conversations." I turned it on, rewound the tape until the whining stopped and one reel was empty, and pressed two knobs simultaneously. The reels revolved in slow motion. I handed a microphone, which was on a tripod to Herby Johnson. "Today, we are going to start by having each student introduce himself and say a little about himself. Herbie, start us off," I said, winking at Herbie.

Herbie's green-brown eyes widened and lit up with the indicator lights on the play dial. All eyes focused on the silver microphone. Herby was dumbstruck. Grunt's upturned nose twitched; he stared at the tape recorder as if it were a keg of dynamite.

"C'mon, Herb, say something," Ludwig urged.

"Yeah, don't be chicken, Herbie . . . or are you scared of listening to your voice?" "Shut up, you assholes," Herbie said, jabbing Hooks with the mike.

"No horseplay with the equipment, guys," I ordered, taking the mike from Herby. "You try it, Johnson." I gave him the mike. "Tell us your name."

"Johnson."

"How old are you?"

"Thirteen."

"How long have you been at Camp?"

"Since April twenty-first."

"What's your favorite activity here?

"Planning escapes," he joked.

"See, that wasn't so bad," I said, running my thumb and index finger over my thick mustache which concealed the upper lip. "Let's play it back. Ever heard your voice the way other people hear it, Johnson?"

I pressed new buttons and the reels whined backward and stopped. Now all the eyes stared at the two built-in speakers in the box. Static crackled and then my voice: "Today is Saturday . . ." My voice was the same but a little flatter. "C'mon, Herbie, say something," Ludwig repeated. "Yeah, don't be a chicken . . . or are you scared of your voice." "Shut up, you assholes," said the recorder in Herbie's voice. "No horseplay . . ." ordered the tape recorder. "Shit, let Green go first. He's got a big fucking mouth," blared the speaker.

Hooks gasped, putting a hand to his thin lip.

"Listen to Hook cussing."

"That doesn't sound like me," Hook blurted.

The group settled down, managing to control their outbursts until each speaker finished his presentation so each could listen to the playback of his voice.

During the remaining afternoon, the boys painted their rooms and cleaned up the mess. I imposed one rule about colors: don't mix them within the same wall. Two-toning one's room was okay. When boys entered the mess hall for the evening meal, they smelled of piney turpentine and had splotches of paint on their county-issued clothing and heavy clodhoppers, drawing considerable attention.

It was my impression that after three months, including emergency dorm duty in B and C, when Dorms filled in for staff taking sick leave, dorm supervision procedures were well established and informal. In practice, DPOs oversaw and meted out discipline. Raising staff and student expectations simultaneously was paramount. Ironically, it was customary to assign a newly hired DPO to the lowest-populated dorm to learn procedures. In my professional view, cracking the institutional culture strategically, A-Dorm was the place to begin.

Adopting core *Summerhill* principles was made easier by the presence of empathic, liberal arts-educated staff and enlightened Probation Department administration that rejected the traditional authoritarian juvenile corrections model. In October 1966, Camp Kilpatrick's program appeared conducive to experimenting with radical treatment reform such as student self-government electing a student mayor, paid jobs, a hierarchical student job program, self-prescriptive recreation program, the right to play sports or be left alone, and off-campus "honor system" allowing personal clothes.

I presented the *Summerhill* idea at a general staff meeting observed by Director Kirkland and the camp's consulting clinical psychologist, who spent most of his bi-weekly consulting day shoring up staff morale playing hoops. I outlined a 7-10-day orientation program combining social work group work skills and the *Summerhill Education* model. While unable to invite A.S., Neil, himself to conduct a colloquium, I managed to invite a young alum of his tutelage at *Summerhill* boarding school in England. The 15-year-old *Summerhill* alum lived next to my home in Seal Beach. Having observed the plenary power DPOs exercise in students' life, including discharge date, I announced as *fait accompli*: A-Dorm was now Camp Kilpatrick's *Intake and Orientation Dorm.* Newbies will be integrated into the camp's general field periods, and mess hall chow schedule, have access to recreation resources, and be eligible for A-Dorm staff-supervised off-campus "treats." After the *Summerhill* alum described his radical self-government school experience, we will follow up by conducting daily group meetings debating camp and dorm issues, promulgating rules governing forthcoming recreation equipment, ping-pong table, and Dorm TV, and sanctions for destroying county and personal property. The immediate plan was to spruce up the Intake dorm in time for the upcoming Sunday's student graduation ceremony and camp's open house honoring the juvenile court volunteer guests that donated TVs for dorms. Lastly, to save time and energy rounding up stragglers, the Director and staff supported proposals to suspend the flag ceremony and ban cuffing mutineers to timeout benches like pilgrims used punishment stocks to humiliate rule breakers.

"Freedom-Not License"

A Summerhillian Explains Self-Government

Summerhill's attendee's mom was a local elementary teacher, and his father was a public defender. They asked me to counsel their 14-year-old son, Sandy's transition from *Summerhill,* a "radical self-governing children's boarding school in England to a conventional public school in ultra-conservative Orange County, California. Sandy's disruptive rebellious behavior and truancy resulted in juvenile court intervention and his parents accepting the judge's private treatment option over county detention. After serious research, his mother selected *Summerhill,* praising A.S. Neil's self-government philosophy and belief that a "child is innately wise and realistic" and should be allowed the freedom to be themselves." (A.S. Neill *Summerhill A Radical Approach to Child. New York: Hart Publishing co. 1960* P.4). Sandy agreed to describe his year *Summerhill* experience to the orientation group.

Scanning my charges locked in a building surrounded by puke-green color concrete walls that reminded me of Fort Lewis's dreary Army barracks, I called a dorm meeting to update them on camp policy changes adopted in the general staff meeting. A-Dorm was officially designated Camp Kilpatrick's *Intake and Orientation program* for new students and would continue to function as a "time out" dorm for B and C dorm students to resolve behavior issues that impeded a positive program adjustment in school and work. Intake students would join other dorms in pre-scheduled field periods, doubling outside leisure time.

"FYI," I said, "I am happy to report that the woodshop has completed constructing the ping-pong table. Paddles, ping-pong balls, and nets are in the admin. But, before the first serve, the group here must establish and post rules regarding sign-up times and consequences of damaging ping-pong equipment. Also, Director Kirkland has confirmed the rumor that the *Women's Juvenile Court Volunteer Organization* has donated a TV for each dorm. Furthermore, he directed staff to spruce up the camp for the open house held in the volunteers' honor set in the following month's Sunday student graduation ceremony."

I closed the meeting, announcing that I was bringing a 15-year-old Orange County adjudicated delinquent and a neighbor to describe his unconventional rehabilitation experience at a unique self-government boarding school in England called *Summerhill*. An education model that I believe had features transferable to Camp Kilpatrick.

I introduced Sandy to a group of newbies on my day off. He quoted Neil's book, "Summerhill is a self-governing school, democratic in form. Everything connected with social, or group, life, including punishment for social offenses, is settled by vote at the Saturday night meeting. Every meeting has a different chairman, and the secretary is a volunteer. Every attendee has a vote including teachers. He described a typical day at Summerhill co-education boarding school. Breakfast for staff and students was from 8:15 to 8:45. Beds were supposed to be made by 9:30 am when posted lesson timetables began in Lab, English, history, geography, mathematics, woodwork, and art and ended at one o'clock. "No pupil is compelled to attend lessons" (p. 12).

Sandy said, "To a *Summerhillian*, freedom means doing what you like so long as you don't interfere with the freedom of others. Kids have the freedom to destroy their property but not license to destroy or steal other people's stuff. Punishment at *Summerhill* is mostly fines related to crime. For example, girls engaging in shenanigans after bedtime must go to bed an hour early two nights. Boys accused of throwing rocks at other boys must cart clods to level the hockey field." Sandy alluded to voluntary counseling sessions that Neil referred to as "private lessons" (PL) or "fireside chats" offered to help new residents in transition from strict conventional to self-government schools.

Neil spoke of gaining a better understanding of dealing with what Neil cited as the "one perennial problem-- *the individual vs. the community"* (SH p.53.)-- that can never be solved." A co-education boarding school like *Summerhill* illustrated this quandary. Boys and girls shared classrooms, leisure time, recreation, and meals but slept in separate beds, and were not given or sold condoms. A school policy Neil maintained was not based on staff morality, but because *Summerhill* would lose its private boarding school license and be shut down. "An individual student may exercise personal freedom boycotting school classes but disrupting classmates' learning environment violates community code. You break somebody else's stuff; you fix it or pay for it. Throw a crazy fit playing ping-pong? You surrender your paddle for three days." Sandy attributed his coming to terms with authority figures' expectations and rules as a strong dose of Summerhill's self-government committed to democratic values.

I took a pair of settled-in newbies awaiting transfer to C-Dorm, off-campus on a scavenger hunt in Malibu, combing the beach for sand dollars and driftwood. We stopped by a *Baskin-Robbins* which gave us empty five-gallon cans to decorate as wastebaskets. A grocery store donated plastic milk cartons for storage bins and bookshelves. I brought tools, wall paint, brushes, poster boards, colored chalk, and chalk from home.

Walking across the recreation field toward the volleyball court, a hawk soaring over the looming horizon, I paused by two boys sitting cross-legged in sand, playing war with an incomplete card deck.

One asked, "Is it true that A-Dorm is getting a ping-pong table?"

I nodded.

"So, what about us?"

"Guess you'll need to convince your DPOs that you all can play without destroying it."

I joined in a volleyball game on *Baby Giant's* losing side against a four-boy C-Dorm team led by DPO W*ildman.*

"Wake up, *Baby Giant*," demanded Jeff Kemper, a.k.a., *Cowboy*, sans western hat, making a gesture to *Baby Giant* to raise his hands and long arms dangling like useless appendages at his side.

"Fuck you, *Cowboy*," *Baby Giant* said.

DPO *Wildman*'s next jump serve came flying hard at an angle, landing between middle blocker *Baby Giant* and me.

"Jesus, what's with you, *Push Broom,*" chided Joe Talmage, a pudgy, short-tempered, pimply-faced blond called "*Garbage*" due to his habit of eating with his mouth open and compulsion to scarf up anything left in his tin chow tray. Victorious, *Wildman's* green eyes scanned the net for a vulnerable spot to nail us in our coffin. The line-drive serve was aimed at *Baby Giant* retreating to the backcourt zone. Luckily, he deflected the ball to our right-side hitter who spiked the ball, momentarily giving us back our dignity until *Wildman* easily returned my sidearm serve in the basket-hands of clueless John Casper, a.k.a., *Ghost,* who forgot he was in this world. Pale skin, silver-blond hair, chubby, consigned to "little kids'' B-Dorm, Casper did not appear to mind his moniker as it was reflective of his surname and habit of never being where he is supposed to be.

Preserving his team's 8-point lead by hogging returns, *Wildman* uncharacteristically gave losers some slack. He served a lob underhanded over our side of the net. My eyes tracked the ball's arc, anxiously thinking about how to return it. I raced to the net, leaped as high as I could, and spiked the ball landing at the feet of *Wildm*an Morris's net man.

"Goddamnit, you guys," said *Wildman*, retrieving the ball and hurling it hard under the net at my knees.

Rotating serves did not help. They'd serve us bloopers, but *Baby Giant* could not time his jumps and swings, and *Garbage* hit the ball as would a girl. No matter. I'd slam it out of reach of his teammates, *Wildman* Morris, hogging his side of the net, always saved it. Soon they lost interest in the one-sided game; players started gawking at other field period activities: peers playing catch with a football, sitting Indian style on the ground, making furrows in the sand with their sneakers, standing in a stupor, staring at animal shape clouds, and lying prone reading paperback novels. Such behavior led the bored observers to conclude: "County Probation officials must go out of their fucking way to recruit the biggest rumdumbs and dipshits at Juvee for *Krazy Kamp*."

"Losers give us five laps," *Wildman* said, offering us no slack.

Jackie Kennedy's Visit

Graduation ceremonies were held on the last Sunday of the month. The mess hall was decorated with white tablecloths, flowers, and bowls of fruit for visiting family members of graduates. All attendees got a Coke and an extra order of dessert. DPOs, peers, and staff gave testimonials on the graduate's program adjustment. They described in colorful terms his first awkward days at camp such as climbing in trees to get attention, hiding in weeds to get out of work or school, temper tantrums, idiosyncrasies like never lacing County boots, aversions toward team sports, and school pranks. Out-of-bounds, or AWOL attempts, and bed-wetting were never broached.

But today's ceremony included special guests: The Chief Juvenile Court Judge and ladies representing *Women Juvenile Court Volunteers.* According to rumors, their spokeswoman would announce their organization's TV donations for the dorms. This event was preceded by countless days Intake Dorm's students spruced up A-Dorm for the scheduled Open House. We looked forward to displaying our hand-crafted sand dollar mobiles suspended from the ceiling, driftwood sculptures, a new Ping-Pong table, students' artwork and brightly colored painted dorm bedroom walls, and bookshelves made in our woodshop.

I sat with a group of newbies at the mess hall when A.D. Racetti led a small group of dignitaries through the door to the back of the room. One of the high-stepping ladies dressed in Sunday best wore a pillbox hat like one previously seen on Jackie Kennedy. The only male guest was a tall Juvenile Court judge that had gray hair, a tan face, and a long beak. He wore a three-piece dark suit, a white shirt, and a red bow tie. A silver pocket-watch chain fell from a vest buttonhole to a vest pocket. Students were dressed in clean County-issued denim shirts, jeans, and clodhoppers. In contrast to normal shifts, DPOs and Rec leaders left their T-shirts, sweat jerseys, and wrinkled shirts in staff lockers or at home.

Guests included parents sharing a table with graduating sons. As was customary, ice cream dessert included an ice-cold bottle of *Coca-Cola* for all. Marking the occasion, a table occupied by a graduate was adorned with a white linen tablecloth, and a single red rose in a vase. Following emotional counselor and peer testimonials depicting graduates' dramatic trials and tribulations, the plan was to give the guests a camp tour starting at Intake Dorm.

That day, Popeye went on and on about his graduate, Miguel Yanzee, applying his Navajo Nation Indian tracking skills to lead Tippy Toes and a group of "lost hikers" back to camp after dark. The camp medic had to administer first aid to the boys cut from crawling through the thickets. Since that incident, any staff leading a hike in surrounding hills had to have completed a compass-reading course and be issued canteens and a flashlight.

At a table in the middle of the crowded mess hall sat Joe Talmage -a.k.a., *Garbage.* He was hunched over the third slice of Neapolitan rainbow ice cream, an orange-pink stream dripping from his chin forming a goatee. Eating with his mouth open, he said, "Wish they'd git, so the camp can get back to normal."

Henry Ludwig, a crew-cut muddy-blond, compulsive14-year-old table-mate, who spoke with a German accent, and was forever repulsed by Talmage's glutinous mumblings said, "Jest give him a trough of food and he's in heaven."

Cowboy joined in, "Never have to worry about *Gar...bage* goin' AWOL. So fat, he couldn't climb the fucking fence."

"No shit. Needs the county to special order his jeans," Larry Jacobson, a newbie, from Glendale,

"Good thing you can breathe through your nose, *Garbage,*" added *Baby Giant,* sitting within earshot at the next table.

The relentless onslaught of painful jibes was a flame racing a burning fuse connected to a block of C-4. Talmage stood, right hand clutching, shaking an unopened bottle of doom, body leaning backward, assuming grenade throwing position, left arm saluting sky, eyes aflame, in-room hushing silence suddenly interrupted by young voices shouting in unison, "Hit the deck!"

Instinctively as if Army trained, I joined a hundred boys and DPOs hunkering down under tables, thanking the gods. Thanking the gods, Talmage's rage propelled the glass missile at bullet speed in a straight trajectory across the room, exploding a thousand shards and brown liquid of disillusionment above the unwary heads of child benefactors. A foot lower and a few yards wider, the bottle would have beaned Jackie Kennedy and imploded hope of camp reforms.

"Jesus Christ!" *Popeye* screamed, yanking the child culprit away from the table and whisking him out the mess hall to A-Dorm.

I apologetically retrieved Jackie Kennedy's purse and fur cape as she anxiously fingered her hair for glass. The other female do-gooders, holding their hands to their cheeks, looked at their friend's rumpled hair, likely thanking the gods that it wasn't them in the line of fire.

"So much for the promised land," DPO *Hippie* whispered, directing kids back to the dorms. The movement was more orderly than usual as if everyone had reached a breaking point and did not wish for another disturbance. This was also the way for the general population to show the Intake kids that came across as cocky and as having special privileges how to act in public society. Most students understood that the guests did not drive up the mountain to check out their welfare. All the hoopla appeared directed toward the Intake program which, at that moment, I felt was in jeopardy.

Talmage was restricted to A-Dorm pending a full staff review chaired by the Camp Director, consulting shrink, and Talmage's DPO counselor. The review was scheduled for midweek, giving me a few days to reflect on strategies to preserve the Intake program gains and hold the provocateurs of the Coke-throwing debacle accountable.

Talmage had no history of physical violence, save smashing his models that turned out less than perfect after they were finished, or throwing cards after losing hearts, gin, or poker betting candy. He collected car magazines and exhibited a low frustration tolerance in woodshop projects, appeared self-conscious about his severe case of acne which peers ignored, but not his overweight and repulsive eating habits that he appeared to flaunt attracting attention. He was the child of an absconding father and a mom working as a geriatric nurse in a nursing home in Torrance. On the outs, he was inclined to backtalk male authority figures, skip school, and liked hanging out at neighborhood auto repair garages. His inconsistent test scores and borderline grades rewarded him with social promotions. Since camp admission in July 1966. Talmage had received absolutely no camp visitors.

I was curious about the mess hall's uniform drill-like response to hearing, "hit the deck," as if throwing a coke-bottle grenade was not uncommon. Cokes were a rare treat reserved only for "graduation" on a Sunday. Indeed, only one such incident was reported previously, a full month before I started at Kilpatrick. The thrower missed a KP that deliberately passed him over when Cokes were handed out. The thrower lost his seat on the Friday night movie bus and was demoted from "office helper" to "sweeper."

I spent the shift meeting with the student provocateurs: John-*Baby Giant* Blake, Jeff-*Cowboy*-Kemper, Larry Jacobson, and Henry Ludwig privately and together with their counselors. I asked each to speculate on Talmage's fate if the hurled coke grenade injured the guest rather than missed.

"Would it impact how camp staff and victims weigh the significance of individual students' role in triggering the incident, and the punishment to follow?"

I wanted to scream, "Damn right, knuckleheads! Would Talmage be at YA, if the woman volunteer bearing gifts for kids here was facing plastic surgery after being disfigured by the exploding glass?" Regarding injury," I continued, "Fate was kind. But your behavior has serious consequences. Your actions jeopardized program progress and placed Talmage's student status here at risk. Where do we go from here?"

Following a long pause, I distributed a blank sheet of paper and a pencil. I spelled out the lady court volunteer's name. I instructed each to compose and sign a letter of apology to her and the other volunteers. I emphasized, "Your letter of apology will not absolve you, but provide some mitigating circumstances to give Talmage a second chance at Camp Kilpatrick in the victims' eyes and our staff."

I discussed his case with fellow DPOs that will deliberate on his fate at the staff review. Had the coke bottle hit and injured the lady guest, Talmage would be history, sitting in the juvenile hall awaiting transfer to secured California Youth Authority's setting. In his defense, there was no evidence of premeditation or intent to cause bodily harm. Talmage's loss of impulse control was triggered by his adjudicated delinquent peers. Talmage got his second chance.

I took a week off.

Who's Krazy?

Upon return, I found the dorm in shambles: a missing ping pong paddle, sand dollar mobiles strewn in pieces, scattered trash, overflowing wastebaskets, transistor radio blaring static, and puzzle pieces s left on tables, *Baby Giant* on restriction confined umpteenth time in Intake dorm for horse-playing his way out of class to escape earthlings in a spaceship to Mars in another science fiction book his mom gave him by the shoebox. Adding to my aggravation was the sight of two newbies, suspended from field period for "throwing crazies" climbing trees and scaling boundary fences, now playing dominoes in the control center with a newly- clueless- hired DPO. Time for a "Reality" check.

The next morning, without explanation, I instructed the newbies, the Coke grenade thrower, and provocateurs to dress in personals and get on the *Camp Vernon L. Kilpatrick* school bus.

Two hours later, piloted by Rec Leader, KC and me as navigator, the *Krazy Kamp* bus pulled into the expansive campus of *Camarillo State Hospital*. With the support of fellow MSW social worker Jack Taub who screened Camp Kilpatrick candidates, the approval of our camp director, *Butterflyman* himself, and *Camarillo State Hospital* officials, I arranged an unprecedented field trip to the state's treatment program for mentally ill adolescents. The purpose was to re-teach our students that Camp Kilpatrick was not L. A. County Probation Department's camp program for "crazy" delinquents.

In the first bleak dayroom, we saw mostly white Kilpatrick-age boys dressed in dirty pajamas and bathrobes, sitting on the floor, their eyes glued to the cartoon images on TVs suspended from the ceiling. Others played with clay, molding prehistoric animals, scribbling with crayons, concentrating on making paper airplanes, perusing comic books and magazines, or sleeping soundly in chairs. A Black boy that appeared no older than 10, wearing a red cape and shorts, sat on top of a desk flapping his arms like a fledgling, and said, "I wonder boy. I wonder boy." A biracial early teen boy rocking in a straight-back chair, his medicated eyes glowing dark marbles, reminded me of my mother's eyes as she madly pumped her legs as if counting down to launch into the unknown. I shuddered with the memory.

According to our guide, the next ward was reserved for the "most serious cases." A pale-faced kid in underwear standing by the door was plucking his hair out. We observed pajama-clad residents, sitting on the floor in their urine, with vacant eyes oblivious to intruders, tongues hanging out like panting hounds after chasing critters. Others remained standing, limbs frozen in various contorted positions like abandoned mannequins or slow-motion bodies miming gibberish. A husky Black male attendant in an immaculate white uniform pointed at two youthful figures curled on the floor, and said, "those patients were in a state called catatonia. Their minds are gone." I watched *Baby Giant* averting his eyes away from an obese Hispanic teen half-standing, compulsively tying and untying his bathrobe, exposing his naked body to anyone walking by.

The word, "crazy" connotes many meanings and images. The goal of the field trip was to help Kilpatrick's adjudicated adolescent outliers better differentiate between themselves and their institutionalized counterparts the courts classified as mentally ill. Sadly, regardless of the cause of the disease (organic, chemical, physical, environmental), *free will,* as it applies to those diagnosed as mentally ill, is excised from the equation due to society's overriding ignorance of human behavior and its complexity. One may have argued that the ongoing mission of a free society lies in the endless pursuit to identify the psychological, organic, political, and environmental attributes and impediments that govern individual mental capacity to exercise free will and set appropriate corresponding accountability standards for violating norms, norms established by law, custom and culture and nature.

The objective of *the Camarillo State Hospital* field trip was **not** a lesson in empathy for children. Personal and professional experience taught me that children generally have more empathy than adults. Kilpatrick residents, Roosevelt Park playmates, St. Agatha and Scioto Country Club peers, and underprivileged Watts kids did not support bullies, mock bedwetters, handicapped or the disfigured, mother-talk orphans, shared not hoard. Human development child experts agree, which I concur, that children are typically two or three years ahead of parents. Curious as cats, children naturally seek out information on sex like my cousin Jane and I explored images of a womb, penis, boobs, and swimming zygotes in her dad's medical books. Kids experimenting with drugs, liquor, and smoking were commonplace along with engaging in high-risk behavior like trespassing government property and moonshiners stills, occasional larceny, hotwiring a friend's car or a boat, vandalizing property such as blowing up commodes and mailboxes with cherry bombs, and cheating on school tests.

Flak Jacket

A dozen boys gathered in the Intake Dorm's dayroom to hear their voices, replaying their reactions to a visit to *Camarillo State Hospital* on a tape recorder.

"So, will anyone venture to describe the differences between the boys we visited at *Camarillo State Hospital* and Camp Kilpatrick?" I asked turning on the *Sony* tape recorder.

"Those people are loonies," Green said.

"What's that mean?" KC asked.

Cowboy leaned forward, his bright-blue eyes looming large, and said, "Nutzo."

"Acting weird; sitting around quiet like a ghost or pretending to be something you're not, like that boy in a cape," *Baby Giant* said.

"That's only partially accurate, Blake," I said. "That patient wasn't pretending or play-acting. In his mind, he was `Wonder Boy` and hospitalized because he cannot distinguish fantasy from reality."

"That's creepy," Ludwig said.

"Do they know they're wacko?" Green asked.

Taking a deep breath, I said,

"Those are individuals that doctors diagnosed as full-blown **psychotics** because the patient believes his off-the-wall behavior is normal like Wonder Boy's belief about himself. A psychiatrist professor described a patient that had the delusion that he was a cow and refused to eat for days; until the therapist told the patient that as a cow he must eat to be fattened up for butchering. If the patient at *Camarillo* that called himself, `Wonder Boy` believed he can fly and found himself standing on the roof of a tall building and leaped to his death, the lethal fall and behavior would **not** be classified as a suicide because his conscious intent was to **fly**, not die.

Others may imagine hearing strange voices, and exhibit symptoms of catatonia that involve extreme social withdrawal, confusion or lack of communication, agitation, or nervousness, which may endanger their own lives in the future. But the treatment of mental illness does not always warrant institutionalization any more than all physical illness requires hospitalization.

What is important for all of you to understand is that individuals may exhibit various degrees of mental illness or whacko behavior and remain free until they hurt someone or become a danger to themselves. Whereas Kilpatrick kids are here because the courts have adjudicated you all as mentally sane juvenile incorrigibles that reject school, ion, and learning social norms that prepare youth for responsible adult citizenship. So, while you are here, we want you to wear a **learning shirt** in class. The law requires it until age 16. In addition, you will find an imaginary **flak jacket** in your footlocker that will help you survive here, and more importantly, survive on the "outs" by fending off the slings and arrows encountered in everyday human interaction. Like a soldier pursuing a mission or athlete seeking victory, one must be prepared mentally and physically. Imagine wearing a flak jacket with three pockets each containing an essential item of a survival kit.

Pocket No. 1 **Perception** of objective; No. 2 **Planning;** Pocket No. 3, **Execution**. First, understand the primary goal e.g., scaling Mt. Everest; Secondly, prioritize the logical steps to achieve the goal. And thirdly, steps in the execution process: setbacks and recovery. Despite what one hears, according to my kin that raised me, there are no natural winners. Achievement comes from opportunity, self-discipline, and hard work."

I was drilling holes in a new batch of sand dollars at my desk in the dorm's control center on the first day of Christmas week. The beginning of twenty-four-hour, school-free days when eight DPOs and REC leaders work overtime catering to the insatiable needs of 90 restless, rambunctious, homesick incorrigible boys. It was a demanding period of coordinating off-campus excursions: fishing at Santa Monica pier, packing picnic lunches, fishing tackle, flip-flops, and beach towels. And organizing intramural campus scrimmages, board games, model building with Testor's glue, out-of-bounds hikes, and squeezed-in family counseling sessions.

Tippy Toes was looking into the mirror combing his hair, killing time until we released our brood to the field. Reacting to the sound of somebody pounding on the dorm door, he walked on the tips of his sneakers to the control center and retrieved the large brass key on the desk, then went to the window by the solid metal door to see who was outside. "Shit, more visitors." He turned around, and instructed the nearest newbie, "Start policing up. We got citizens," coming through."

I heard a female voice and a commotion in the control center.

"Set it down here," said *Tippy Toes* to *Popeye* Benny and a C-Dorm kid carrying a large cardboard box labeled *Zenith TV*. Wide eyes, *Tippy Toes* continued, "Weren't you . . ?"

With laughter in her voice the woman replied, "Yes, I forgot to duck."

Rising, dropping the drill, discombobulated, I found myself standing arm's length from lady bountiful, a tall middle-aged, attractive short hair brunette woman without a shard of glass or pillbox hat. She was dressed in an impeccable brown corduroy suit, blue shirt, and handmade leather boots, an image right out of *Abercrombie & Fitch's* fall catalog.

"Sorry about that, ma'am," I said, as Benny attacked the box and set the shiny twenty-one-inch *Zenith* TV with push-button channel dials on the control center counter facing the dayroom.

"Gloria Lambert," she said, extending her hand. "They won't work until you install the antenna on the roof. Also brought some art supplies." She held out a heavy bag.

New Year 1967

"My little monsters are hanging around the control center like flies on shit, fighting over shows to watch," DPO *Hippie* complained during a morning coffee break on New Year's Day, 1967.

"So, hold group meetings to vote on programs they want to watch," I replied.

DPO *Choo, Choo* sighed, "All my kids want to watch is *Hogan's Heroes* or *Batman*."

Meanwhile, as my wife was standing on the curb of a crowded Pasadena Street with a friend, smelling the roses, and watching the *Rose Parade,* I was staring out a window at homesick *Krazy Kampers* engaged in horseplay during the morning field period. It took me back to St. Agatha's unstructured recess days when nuns stayed in the shadows, fingering rosary beads, eyeing boys playing football or soccer on the grass, and girls standing around in cliques trading secrets and gossiping. St. Agatha's recess was an escape from hours of desk confinement, listening to boring catechism, rote learning, regurgitation of dates, names of saints, heroes, and villains, and standing erect facing a blackboard, hoping you do not get a boner diagramming a sentence.

I appreciated Jack Taub's clinical social worker skill in identifying a "delinquent" sub-population of "incorrigibles" struggling to find their "creative sweet spot" that sparks the beginning of one's existential journey. The unconscious "sweet spot" is heretofore repressed, hidden, or lost in the labyrinth of societal regimental conformity or parental psychological imprisonment. Fortuitously, the county Probation Department recruited Kilpatrick staff whose *Inner Child of the Past* led them as adults to advocations and careers in education, psychology, arts, writing, photography, and human services. This shared inner child dynamic underscoring A.S. Neil's belief in children's right to play explained our DPOs' default to peaceful over violent conflict resolution with volatile troubled youth.

My colleagues shared the view that our kids were casualties of broken homes and California's rigid compulsory education policy. In the 60s, status offender statistics were underreported and or counted in juvenile crime reports. Many kids were junior petty criminals like the writer who got a pass by fate or his Aunt Helen's social station. In California, violent juvenile offenders were committed to the California *Youth Authority. S*tatus offenders generally were a county responsibility. Notwithstanding California juvenile and child welfare laws and related philanthropic resources, thousands of underage runaway youths found themselves surviving on the street alone. Many joined the "flower children" flocking to *Haight Ashbury* and sought refuge in counterculture communes. Others, sadly declared, "missing children," became victims of sex traffickers or drug dealers.

Saturday through Monday the Senior DPO schedule demanded constructive program planning instead of school classes. Security and activities were dependent on the cooperation and coordination of an Assistant Director, eight DPOs, and two recreation leaders. The kitchen operation was supervised by adjacent forestry camp personnel that shared the mess hall. Three scheduled meals a day including sack lunch orders for off-campus outings. A DPO was encouraged to take his eligible caseload dressed in civvies to Santa Monica Beach or town for a few hours. No policy dictated staff supervised off-campus excursions. However, most DPOs engaged in group projects such as model building, off-campus hikes, sports, picnics, and family counseling sessions.

Boys that completed the Intake program more and more served as reform cadres to change program expectations of other dorms. By spring 1967 the goal of transforming A-Dorm from a disciplinary and wayward station for new admissions to an Intake and Orientation program was well underway. The shock of transferring from the more intimate and intense intake of "promised land" to a large, regimented, and lethargic regular dorm motivated students to make demands for improvement. Other dorms posted rules, a daily TV schedule, and art drawings covered the walls. The ping-pong table was liberated. Ex-newbies successfully hounded dorm DPOs to conduct group meetings and vote on TV programs. Scheduled Field Periods offered students the option to remain in a dorm, watching TV, playing ping pong, cards or reading if a DPO was present. Destroying county property significantly decreased.

Thanks to an enlightened Central Administration, staff civil service status (DPOs) were not subject to conservative backlash against DPO's casual dress code and a radical treatment model based on A.S. Neil's philosophy and belief in a child's innate freedom without a license. This allowed them to enact a program that included elements of children's self-government, democratic rules, and consequences.

Lance

Hellbent on escape, a jockey's height, freckled face, rust-color flattop, beady eyes of a caged bird, 13-year-old Lance Penney cased his new home the moment he arrived at the infamous *Krazy Kamp*. According to the case file, Lance was a poster juvenile status offender. He held his school's truancy record and was picked up by police while wandering the streets. But Child Welfare found no evidence of parental neglect or abuse warranting foster care placement. Lance lived with his mother and four siblings in a three-bedroom apartment complex that she managed, located in a working-class neighborhood in Alhambra. His father was a career Army sergeant stationed in Germany. School test scores placed him in the 75th and 80th percentile in language and math respectively. Skidding by on Cs and Ds, Jack Taub felt Camp Kilpatrick's unique population of young underachievers and social outliers may awaken Lance's latent attributes. Attributes that will abate his rebellion and open up with adult authority figures.

Sitting at a picnic table in the visiting area during the field period, I introduced myself to Lance as his DPO counselor after he completed orientation and was transferred to C-Dorm which housed younger boys. In the initial counseling session, he expressed no interest in sports or hobbies and did not explain missing school.

"Somedays, I'd "head off to class, but just kept going," he told me. He liked, "sneaking into movies to see western and war movies." At home, his brothers monopolized the TV to watch sports."

I asked, "When was the last time you saw or spoke to your mother?"

He said, "My mom's job keeps her too busy." He complained that his brothers took over his bedroom space when he was away in juvee. "Put my stuff in boxes like I was fucking dead."

I asked him to recall his favorite day or experience with his family. He cited a family vacation at *Veterans Park at Redondo Beach* the previous year to celebrate his dad's birthday while on military leave and promotion to Master Sergeant. He said, "My mom got so sunburnt, that my dad rented a canopy. The rest of us visited old shipwrecks and souvenir shops."

I informed him that he was one of four boys in my caseload and that it was my practice every other Sunday afternoon when I'm on duty to take individuals to *Santa Monica Pier* and the beach. Maybe we'd go fishing, eat pizza and ice cream, or just hang loose walking around. The only requirements were not being on disciplinary restriction for behavior problems or failing a class. I asked if he would like to talk with his mom if I called her from the Director's office empty on weekends. He nodded.

Notwithstanding program enrichments, monthly boundary fence violations remained immutable. Absent Without Leave (AWOL) from a relatively open correction setting with a flimsy wire boundary fence more symbolic than a deterrent, hardly conjured up the image of escaped delinquents endangering the public. In the final analysis, Camp Kilpatrick was the placement of last resort for the county's youngest truant and incorrigible children.

Given our students' history to abscond, and children's adventurous nature, we believed that it was imperative to be transparent and realistic about the geographical perils of running away. Regarding race and ethnicity, in the mid-60s, Black and Hispanic youth committed to Camp Kilpatrick rarely had a history of gang affiliation, and thus were unrepresentative of Los Angeles County juvenile offenders. Indeed, Camp Kilpatrick's children of color were socio-economically like Children-at-risk in Watts.

Lance called home from the Director's office available on weekends for family counseling sessions and phone contacts. He told his mother that he was transferred from Intake to C-Dorm and needed new clothes to wear off-campus. During our first session, I explained the program and encouraged her to visit. Her tone was distant and devoid of relief normally expressed by mothers separated from children. She said her other sons, "towed the line," and that Lance's school problems coincided with her husband's absence. "When Lance's dad was a recruiting sergeant here, he gave Lance the attention his brothers got." She said her job "kept them from living in a trailer."

This, unfortunately, was a valid statement.

In 1966, the Army annual housing allowance for enlisted E-5 rank with six dependents was $5000, less than half ($10,500) the salary Los Angeles County paid a Senior Deputy Probation Officer, Step III.

After the conversation with her, I visited Lance. I told him that if he made it through the first week without being gigged, I would take him to the beach with Randall Cunningham who was counting his days to graduation. Lance needed, I thought, *something to look forward to.*

"Fasten your seat belts," I said, keeping my promise, Lance sliding in the *Porsche* Cabriolet's top-down cramped beltless jump seat, behind Randall who was occupying the front passenger's seat. My right hand grasped the vibrating plastic gear knob shifting toward freedom's switchback descent powered by screaming Elite 356 air-cooled pushrod OHV flat-four precision engine muffling sounds of tires spitting gravel and heavy breathing,

After we combed the beach for seashells and driftwood, we stopped at a fast-food drive-in near the Malibu hills. The parking lot was crowded with two subpopulations of California beach tribes: Custom 1950s cars and woody station wagons laden with surfboards, surrounded by Coppertone tan bikini-clad nymphs hanging on long arms of sculpted bodies of cocky male surfers wearing baggy trunks exposing cracks of pale asses. Nearby cafes were loaded with famished teens, their mouths sucking creamy malts through bent straws, sharing salty fries, and lustful thoughts. Nearby, a herd of dusty *Harley-Davidson* motorcycles parked side-by-side awaiting their riders' return. Riders were easily recognizable as road outlaws: Nazi helmets, untrimmed beards on grim reaper faces, black leather vests with skulls and crossbones and wing patches, grimy jeans, truck driver chain wallets, black square-toed *Frye* engineer boots with ankle straps, ponytails, dark shades, and tattooed arms. Some straddled their iron bikes with similarly attired female companions leaning on their outlaw shoulders. Others stood in small groups engaging in small talk.

"Those guys are Hell's Angels," Randall warned.

I carefully opened the door to avoid touching the man straddling a classic red color Harley with whitewall road tires and saddlebags.

"Whatdafuckya thinks ya doin', parking in my spot," the Harley driver said, climbing off his bike, tilting his head back as he spoke.

Exiting the car, I closed the door, surveying the parking lot, and gave a friendly grin. "Sorry, last spot." As I turned around to ask the boys for their food order, a loud crunching sound like colliding bumper cars at the amusement park cracked the air.

"Hey, motherfucker!" the biker said, removing his helmet and setting it on the bike's seat across the Porsche's driver's side door with two crescent-shaped saucer-sized dents in the panel.

"Let's get outta here, Mister Wheeler," Lance said, as other bikers wandered toward our car.

Seeing the sun reflecting dents on ebony-color steel, I stiffened my shoulders and twisted my head, as if I had a crick in his neck. "Figure it'll take about fifty bucks to make that door right," I said, eyeballing the stocky biker with uneven salt-and-pepper beard masking his face, but not his mean dark eyes.

"What makes you think you even get fifty cents from me?" He grinned, inserting sausage-size thumbs in his belt loop.

"Fuck this shit!" growled another biker flexing biceps matching Popeye Benny's pumping iron, exposed outside a black leather wool-lined vest breaking out with WWII German military patches.

"Leave this little fag to me, boys," he added, looking at gathering biker buddies for approval.

I scanned the parking area and highway for a sign of cops, no cavalry in sight. Fuck. I was surrounded by cretins from hell, and a growing ring of surfer rubberneckers waiting to see the carnage surely to come. The army trained how to kill in a single thrust of a bayonet, slit throat with a commando knife. But now, weaponless as a babe in the woods in a no-holds-bar street fight against a dozen fists and steel-toe boots, the singular hope was a lethal blow to stop the pain. I glanced at Lance and Randall hunkering down, hands covering their eyes. Directing my attention to the bull-neck leader's laced fingers laden with flashy silver cross skull-bone rings, cracking his knuckles in sync with my racing heart. I envisioned meaty hands grabbing the back of my shirt collar, lifting me off my feet, hurling my dangling body into a metal hog, knocking over a row of Harleys falling like dominoes. The sound of glass, crunching plastic, and steel incites the vengeful owners stomping my curled body to a heap of flesh and bones as passersby listening to rock & roll music on car radios, take no more notice than seagulls hovering for leftovers.

But the next voice I heard inching back to my car was not gulling birds or sirens of police cruisers. It was the high-pitched raspy female voice behind me.

"He's got the message, Jake!" said a slender, copper-faced woman with long, braided blond hair, tight jeans, and Levi's jacket decorated with a large patch showing a skull-face angel on a white horse. Her radiant blue eyes glared at the stocky biker like an excited hawk. My greatest fear was the bikers preventing me from returning my charges to camp.

"Sorry guys," I said, snapping the seat belt, welcoming the sound of Porsche's screaming engine.

Out Of Bounds

Student discipline problems generally coincided with DPOs' off-duty schedule (106 of 168-hour week). The common denominator was analogous to behavior problems at home stemming from parents' broken promises and the absence of a male role model. Like a ball ricocheting inside a pinball machine in "tilt" mode. "Runaways" regressed to "Out-of-bound" impulses once their DPO was off-site.

Lance continuously tested these norms. Invariably, upon my return to duty, I found Lance restricted in the Intake Dorm for boundary violation midweek during the field period. He would hide in the brush most of the day, then eventually wander back to the dorm or Administration building. The Intake Dorm appeared more and more in Lance's camp experience during weekdays. We briefly considered "Intake dorm schooling" to prevent him from failing school. But this was impractical and counter-rehabilitative. Neither his mother nor siblings responded to repeated requests to visit him. No record of his father's current address was in his file. Lance was essentially an orphan with a "paper" family.

A month later during my scheduled off duty, Lance made it down the mountain to the Pacific Coast Highway wearing his county-issued clothing. Santa Monica Police picked him up, thumbing North. Three days later without my knowledge, Lance was transferred to the California Youth Authority facility in Whittier. Whatever self-confidence accrued in counseling troubled youth attributable to graduate school's training in cognitive and group psychotherapy, personal verbal skills imparted by strangers, and significant others, and experiences bestowed by fate in my unique cross-cultural childhood, quickly succumbed to a feeling of profound failure.

Secured correctional institutions are designed to intimidate inmates and visitors: Stark architecture, sterile rooms with locked doors, guards, drab-color walls, wire mesh screen windows, and bars that are anathema to natural light. No pets, flowers, plants, or any semblance of nature were allowed. To allay my guilt, I made a surprise visit on the first day off duty.

Lance walked into the empty visiting room dressed in a clean-pressed YA-issued khaki color uniform, and boots. His demeanor reminded me of dragging my feet to retrieve a failing report card in third grade from the priest at St. Christopher Elementary School: cast down eyes, shoulders slumped. We shook hands like two losers, and sank into wooden chairs, facing each other across a narrow table. Blaming myself for not stopping the summary transfer of a barely 14-year-old wayward youth to a children's prison, I apologized for the weeks-delay in tracking him down in the county/state juvenile corrections complex working at cross purposes. In a tone of indignation against the process, I outlined my strategy for appealing his case, on the condition Lance committed to staying within bounds at *Krazy Kamp* if readmitted.

Reading my mind and tone, Lance rose from the chair, reached across the table, both hands clasping my hand, and said, "No, No, Mr. Wheeler, this is not your fault. For the first time, I realize I need walls to keep me from running away from my problems. Walls force me to stop running and look at myself."

Flummoxed by Lance's reaction and words, "Jesus Christ," I said to myself. On the punishment scale, incarceration was the most punitive and oppressive criminal sanction, not a tool of introspection. Therapeutically, Lance's words were evidence of "psychological insight" that according to Wikipedia "occurs when a solution to a problem presents itself quickly and without warning. It is the sudden discovery of the correct solution following incorrect attempts based on trial and error."

After a reflective pause, I praised Lance for taking a giant step forward. I reminded him of his academic potential, demonstrated in not failing a grade despite missing so much school "on the outs" and at camp. I stressed the importance of identifying an area of interest, setting a goal, and analyzing achievements and mistakes. Alluding to my personal history including running away at age 16 from an authoritarian and abusive uncle not of my blood to avoid a violent confrontation and the adventure of hitching back to Ohio to the aunt who sent me to California, I shared with him how I made plans and carried them out. My preferred plan, plan A, was to sign up for high school football and reunite with old friends. If my aunt did not take me back, plan B was to enlist in the Army four months later when I was 17 if she was willing to sign the papers. The Army offered education opportunities to complete a GED and college later with the GI Bill. Lucky for me my aunt in Ohio believed strongly in education– both her children attended and graduated from college– so, she took me back into her home, and let me choose the high school and eventually college.

In contrast to my uprooted childhood experience, rebellion, escape, and rescue, Lance reacted to his family's inattention by engaging in self-exile. I encouraged him to "fight for your space" in the family, starting with a letter to his mother sharing his feelings of loss of place, his understanding of her burden as a breadwinner, and his specific plans to contribute to the homefront and succeed in school. I told him how I ran away from an oppressive home environment by returning to my roots and setting realistic personal goals in sports, and education.

I hoped that he could see me as an ally rather than an oppressor.

Century Plaza: Make Love, Not War

Three days a week, I missed seeing TV evenings broadcast of the Army's "Five O'clock Follies," Pentagon lies on the Vietnam war. Kilpatrick staff were past the draft age. Leisure time was rare on weekend coverage. Completing court progress reports, conducting individual, group, and family counseling duties and crisis intervention infringed on our sleep schedules. Fortunately, our kids and staff were at chow during the evening news. Our kids had enough to worry about without seeing graphic war images.

At home, there were no distractions to seeing living color-TV images of flag-draped coffins unloading at Dover and reading the five hundred KIA monthly names of the fallen, scrolling across the TV screen like surreal movie credits of the country's anti-communist crusade and juxtaposed to anti-Vietnam war street demonstrators and sounds of burglar taps, gun salutes, hunched mothers clutching folded flags at funerals in small-town America. Despondently I watched TV protests, reeling from narrowly escaping my arrest and assault by a thousand of Los Angeles finest wielding nightsticks at over 10,000 anti-Vietnam war protestors while marching on *Avenue of the Stars* near the *Century Plaza Hotel* on June 23, 1967. A cross-section of Angelenos, Southern California tourists, college students, mothers, children, grandparents, and Vietnam war vets marched, brandishing homemade placards: a cartoon caricature of President Johnson as Uncle Sam pointing a pistol, saying, **"I Want You,"** motherfucker, an enlarged photograph of a naked girl running out of the village, her hair aflame, charred arms reaching skyward, her burnt lips crying for Buddha's mercy unanswered, poster-size images of peace doves-endangered species. Five thousand voices chanted, **"Hey, hey, LBJ, How Many Children Have You Killed Today?"** We delivered our message directly to our audience: well-heeled patrons standing on the balcony of *Century Plaza Hotel* toasting champagne with President Lyndon B. Johnson at a thousand-dollar-a-plate Democratic Party fundraiser. As we marched by, we saw them looking down at America's conscious protesting against the relentless war train that they fueled by lies, racism, and hypocrisy. Suddenly, our loving voices screaming for peace were answered by law & order bullhorns, waves of armed men in dark uniforms and white helmets, swinging nightsticks, shoving, beating, and knocking over the peaceful protestors: civilians, workers, helping the wounded, panic-stricken mothers holding children's hands and pushing baby strollers, wheelchair bound veterans, seniors assisted by canes, nuns, and priests wearing black. Nearly crushed by a wave of fleeing humanity, my wife, a fireman friend, and his wife and I narrowly jumped out of the crowd's way only to be attacked by cops with raised batons. They selected the most vulnerable to beat, cuff, and drag off. Hunching forward, fighting for space, arms protecting head and face, cops corralling us, my friend, a U.S. Coast Guard reservist retrieved a silver City Fireman badge, held it to the cop's face, screaming, "Jesus Christ, Officer, I am a Fireman!" Following his cue, I flashed a Los Angeles County Senior Deputy Probation Officer depicting images of Pomona, the Roman goddess of fruit trees, the symbols of a cross, oil derrick, Spanish galleon, and tuna sculpted in gold relief.

We were released but not without contempt.

Orange Sky

Since TVs were installed in the dorms, Sunday was the most peaceful day of the week because no resident wanted to miss his weekly fix of *Laugh-In.* Tonight, I was covering B-Dorm for the DPO that called in sick. At 8 p.m., duty staff included three DPOs: Benny-C-Dorm, Berryhill, and Henry-Admin on dorm reserve, one rec leader, KC-Intake, and the Senior DPO in charge, filling in for vacationing Assistant Director.

Around approximately 2015 hrs. on Sunday the 15th of October, everyone gathered around the B-Dorm Control Center, eyes locked in a trance on a black & white TV screen with live images of *Laugh-In's* roller-coaster pacing zany characters' antics, and insane skits of slapstick humor. These skits reminded adults and teenagers alike of life in *Krazy Kamp*. Suddenly, halfway through the show, the TV went blank, and the dorm was cast into darkness, causing an uproar. I blindly retrieved a flashlight from the desk and the phone. It was dead.

"All the lights are out!" someone yelled from the door.

I turned on the flashlight, ran to the windows, and saw only a dark field and a faint outline of the Intake Dorm, two football fields away. The floodlights were out, too.

"What's happening, Mister Wheeler?" asked John-, aka Ghost-Casper. No one called me "Push Broom" to my face in intense situations.

"Don't know."

I joined startled boys rushing outside, the pungent scent of ashes invading my nostrils and burning lungs. Eyes burning as if under a tear gas attack that I recognized from my Army days at basic training, I looked upward, forming a visor with one hand to see. "Jesus!" I mumbled under my breath. Bright orange and red streaks were swirling overhead. The sky was raining a trillion tiny, luminous, gold, and silver flakes, all fading like pyrotechnics as they fell to earth. They were followed by a moving river of fiery smoke crawling toward us. The conflagration brought on by a lethal mixture of drought, mountain crosswinds, and man's negligence that consumed all the surrounding kindling, lowered the horizon and transformed the sounds of scurrying birds, foraging nocturnal animals, and humming foliage into an eerie quiet and contemptuous hiss. Other barrages of flashing light followed flakes melting in the molten firmament as if the universe were topsy-turvy and heaven had suddenly become hell. As groups of students of two open dorms and DPOs stood outside frozen, gasping at the fiery spectacle in the sky, the brush landscape bordering the desert basin carved by the county bulldozers for incorrigibles miraculously remained untouched by flames.

In the early weeks, staff and residents never saw me overseeing a big dorm. To them, the Intake Dorm still had the reputation of being lockup disciplinary and the "Baby Dorm." They also heard other DPOs questioning my ability to handle the mass of kids in the large dorms and questioning my commitment to stay in the program. To them, social work was predominantly a female profession as were individual and family counselors. Most social workers and counselors at the time were female. But over 12 months of perseverance beginning with abolishing flag ceremony, granting newbies mess hall and recreation privileges with the general population, dorm makeover, recreation equipment, art, abolishing solitary lockup, group therapy, and field trips eliminated doubters. However, here was my first test for managing large groups of kids. I knew I had to muster every skill I had if we were to stay alive.

The first reaction was the sudden realization I was responsible for ninety-five lives trapped in a mountain wildfire. The absurd irony was that we issued each kid a surplus Forestry Department tin hat and oversize fire jacket the first day at camp, but we never held a fucking fire drill. No official plan existed. I was on my own.

My first thought was to prevent panic by outlining a survival plan. I ordered everyone to return to the dorm, dressed in boots, a tin hat, and a fire jacket, and standby for a headcount. Every single name on the dorm roster would be accounted for. I assigned a runner to carry messages to KC-covering Intake Dorm, DPO Popeye-Benny at C-Dorm. I sent Ludwig to Admin, instructing him to convey to DPOs-Hippie Henry and Choo Choo Berryhill to scavenge the office and rooms for flashlights, fire extinguishers, and transistor radio, then report to me at B-Dorm.

"What we gonna do?" A voice in the dark said.

"From what I can figure, we got a total power out," I said. "Anyone here got a transistor radio?"

"Yes, I got one," a kid said, racing to a footlocker.

Ludwig, DPOs Henry, and Berryhill double-timed from Admin to B-Dorm with three flashlights and a transistor radio.

"Jesus," DPO Henry said, appearing anxious, pointing a flashlight outside.

I was checking off the roster as students donned fire jackets, laced their boots, and filled canteens with water. Ludwig was holding the hand-size radio to his ear dialing for a station.

DPO Berryhill said, "Only news I've gotten comes from the two-bit radio that says there's a wildfire in the Santa Monica Mountains."

"Can't they send somebody to rescue us?" Ludwig said.

"Fires probably closed the roads," I said.

"With these mountain winds, firebreaks probably too narrow to contain the spread," Berryhill added.

"Goddam, I got to call home," Henry said, visibly shaking.

Reacting to the first sign of fear of DPO, our already anxious boys grew even more worried. Children surreptitiously study adult body language, words, tone, facial expression, and stress reactions. They understand their dependence on adult succor, security, and their place in the hierarchy of authority. The Sunday night weekly fix of *Laugh-In* was the apotheosis of *Krazy Kamp's* group culture played out in the adult TV comedy skit depicting America's mid-60s crazy contradictory world of drugs, war, sex, civil strife, and rock & roll. Standing face to face, my eyes locked on DPO Henry's blinking eyes, my hand gripping his arm like a vice, platoon leader's voice speaking forcefully without moving lips, hand sitting him down in desk chair, I said, "**DPO, Henry**, calm fucking down." Then I stood on the desk in the control center, shone the flashlight on my face, and said, "**Quiet,"** pausing for the state of dead silence drilled in newbies on day 1 of orientation declaring an emergency, whereby boys, "freeze" in place. "Listen up, fellow campers, Mother nature is challenging us to the test of our lives, surviving a wildfire until help arrives. Fortunately, except for a few trees in the visiting area, Camp Kilpatrick is the safest piece of county real estate--a desert sand oasis--on the mountaintop. We have the advantage of concrete block buildings, not wood construction, days of non-perishable food at the mess hall, drinking water, and hoses to water down the dorm perimeters. Now is **not** the time to question DPOs' instructions or go out of bounds."

I sent DPO Henry with a flashlight and five B-Dorm boys to check on the food supply, particularly perishables and ice, and hose down any high grass on the interior perimeter of the mess hall. I put DPO Berryhill in charge of B-Dorm while I checked out Intake and C-Dorms.

Walking across the recreation field toward Intake Dorm, I paused and studied the snow-like flakes raining down from the red-orange sky; flakes landing in my cupped hand, instantly turned to powder once rubbed between my fingers. The surreal sky reminded me of smokejumpers in the *Red Skies of Montana,* the movie version of the deadly *Mann Gulch Fire.* The smell of smoke and ashes grew stronger, impairing nasal breathing and burning my lungs. Within seconds, my Levi's jacket was coated in white flakes.

"God, Jerry," KC said, standing in the doorway out of earshot of 10 newbies, lowering his flashlight beam. "Doesn't look good out there. Are we going to evacuate?"

"Negative," I answered in an authoritative tone. "As I see it, KC, the camp's lost all outside communication. We must keep our cool and give our kids a task to focus on and stay busy. First, make sure every boy is wearing boots, has a fire jacket, a tin hat, a bandana to use as a face mask, and a water-filled canteen at the ready. Then take your group to the mess hall with the mission of gathering non-perishable food items like Cokes, fruit, candy bars, crackers etcetera: to last 24 hours. Keep it in the Control Center. I'm going to organize the same, plus a fire brigade at C and B-Dorm. We will keep in communication with runners."

I scanned the newbies mimicking KC snapping on fire jackets and donning World War I-shaped dough boys' helmets, and Buffalo Dog waiting for his master's orders.

Benny, the camp's most popular DPO, was knee-deep in boys standing outside C-Dorm's entrance focusing their attention on their leader's face more than on the ash storm and surreal sky.

"What the hell's going on, Mr. Wheeler?" a kid demanded.

"Hell 's going on," Benny answered matter-of-factly.

"Everyone accounted for?" I said.

"Yes sir," Benny said, handing me a clipboard containing the C-Dorm's roster.

"Great," I said, ordering everyone back inside. Joined by Benny, I outlined the mission of forming four boy groups: "**Group One**: water down the high grass and brush around C-Dorm; **Group Two:** hose down around the gym. **Groups Three and Four**: gather a day's supply of Cokes, fruit, candy bars, crackers etcetera at the mess hall and bring it to the Control Center.

I stepped outside, plucked a patch of ash-covered grass, and threw it into the air, tracking its drift toward Visitors Hill in a beam of a flashlight. I felt the greatest threat-crosswinds was mitigated by cinder block constructed buildings, Visitors Hill, and the camp's basin-shaped mountaintop landscape carved by earthmoving equipment to accommodate the recreation area.

Observing the C-Dorm kids wearing their fire jackets, bandanna kerchiefs, tin hats, and laced boots, I took Benny aside and gave him the pep talk I gave KC, *Hippie*, and *Choo Choo*. I said, "Popeye, give these boys a goddamn sense of mission; a fuckin' adventure."

I was thankful that the camp's current student population had passed through the new Intake orientation program that stressed A.S. Neil's *Summerhill, Freedom Not License;* the imaginary three-pocket *Flak Jacket* survival kit: *Perception, Planning,* and *Execution.* I reflected on the field trip encounter with psychotic adolescent patients that hopefully disabused Kilpatrick students' misperception of "crazy" behavior, the wise clinical tutelage of a child psychiatrist, Dr. Hugh Missildine, and the radical grassroots community organization by Sociologist, Professor John Behling. And lastly, Ohio National Guard's OCS Leadership and Command training and internalized honoring the "officer code" entrusting a leader with his men's welfare.

"Don't think DPO Henry gonna be much help, Mister Wheeler," a B-Dorm student said, as I accelerated through a crowd of boys holding a lit flashlight.

"Explain."

"He looks sick."

"Christ," I said to myself, observing DPO Henry sitting on the floor behind the desk talking to a dead phone.

"Forget the phone, Henry," I said.

"Gotta call my wife, and for help," he reiterated. No expression except for a blank, unemotional stare at the wall crossed his long, tan face.

"Goddamnit, Henry, authorities know we're here. Roads are probably blocked."

Kneeling, face-to-face, I attacked his fingers clutching the phone and setting it on the floor. My hands held his head in a vice, tightening the grip until he established eye contact. In a forceful tone with a raised voice, I said, "Henry, we don't have time for this shit. Count down from three, if you're not on your fucking feet, I'm going to drag your ass to admin. and handcuff you to a commode until we're relieved. Underfuckingstand?"

"Berryhill," I said, directing the man behind me "Hold the fort."

Forty pairs of eyes locked on the Senior DPO in charge, leading a shell-shock soldier arm-in-arm, into the darkness of no-man's-land.

Outside, we ran into KC armed with a flashlight, escorting a small group of newbies wearing dented tin hats, red bandanna facemasks, bulky fire-jackets, pockets bulging with apples, oranges, bottled Coke, and boxes of *Cracker Jacks*, unseen, *Mars,* and *Hershey* bars stuffed in jeans, water-filled canteens drooping on GI pistol belts, and soot-covered boots. "Henry's losing it, KC," I whispered, shoving Henry toward the admin building.

I walked Henry to the staff's sleeping quarters at admin, warning him that I'd handcuff him there if he left the building. Returning to B-Dorm, my mind raced through stages of survival scenarios if wildfire threatened to engulf dorms, causing death by smoke asphyxiation. Our last stand......., mess hall, gym?

"Keep the water hoses going," I shouted. This was a far cry from the days of hiding under my desk in nuclear bomb drills at St. Agatha's elementary school, our "heads between our legs, kissing our asses goodbye."

At 2240 hours, most boys were exhausted, their eyes drooping or sound asleep in their clothes with boots on. Excluding DPO Henry under house arrest, we had four staff for 24-hour coverage of 90 boys in three dorms. Outside, the relentless ash storm continued. No sign of rescue, but we were not panicking. I made the rounds, and all DPOs including me, strode through camp high on adrenaline, projecting leadership and control of the situation. A sense of hope, optimism, and confidence in survival is made easier by the human bond forged in the unique 62-hour adult-child interaction in the unorthodox *Krazy Kamp* setting. Our program that did not impose rules by fear or corporal punishment helped us fight the fire. We saw firsthand how the *Summerhill* philosophy that gave kids the space and security to test themselves in the benevolent milieu of work, study, and play without ridicule or shame was working. The kids confidently worked together to save our asses and our camp.

The next morning, at 0700 hours, my eyes burned from the acrid smoke that permeated the dorms. In the early morning light, the landscape was covered in ashes the color of dirty snow. I paused in the middle of the recreation field and studied the sky. The overlay of reddish-orange clouds was more and more obscured by billows of white smoke drifting toward the sea, and most of the debris seemed to be burning out before it hit the ground. Excited DPOs and kids ran out of the big dorms, soon joined by KC, Buffalo dog, and newbies. I instructed everyone to gather at the mess hall for a general assembly.

Assuming a platoon leader's stance in the middle of the silent room, surveying ninety anxious boys straddling seats of floor-bolted county tables, I took a deep breath, and said with a raised voice, in a joking tone, "Gentlemen, thank you for the orderly movement and attention. It appears mother nature has given us the message; this is not a good time to contemplate attempting AWOL. The fire trickster got us surrounded. However, I am convinced, we are in the safest place on the mountain. We have ample food and water. The crosswinds carrying some of the ambers from the thickets spreading the fire have settled down. More importantly, we're protected by being in this manmade desert mountain top basin with concrete buildings, not wood. And at last look, I saw puffs of white smoke rising on the horizon. A sign that firemen are spraying water to save ranches in the foothills. I think the fire has about run its course."

A loud cheer erupted from campers and DPOs alike. The boys raised their glass bottles of *Coca-Cola*, clinking them against one another and chugging them down, completing a human conveyor belt journey to tastebuds of smiling, dry-throated wildfire survivors with dirty faces. Their chins dripped warm, dark, syrupy, fizzy foam with a taste of cinnamon, vanilla, and a trace of nutmeg. Not even on graduation were the boys so euphonical, a testament to the boys' effort and success in saving *Krazy Kamp.* My body leaning against the table on my elbows, I tried to conceal my fatigue and a pounding headache. The reflection in the mess hall window, matted hair, shadowy beard, told another story; reflected at me was a runaway kid hitchhiking in desert twilight, sleepless mahogany-brown eyes squinting at a *Route 66* road sign. I shook my head and the image faded. Maybe it was just the weird fucking light.

Miraculously, the fire jumped sections of Encinal Canyon Drive leading to the camp's parking area. Patches of grass and flowerbeds in front of the building looked as if hit by a flamethrower: dark gray ashes and burnt sand. But the flames had not penetrated the cinder block building or the metal roof. The County bus, station wagon, and staff vehicles were covered with soot, but the tires were intact. The kids did not mention DPO Henry's breakdown and house arrest. Most of us revisited our fears from that night in silence. I later learned, according to the National Park Service, that on October 15, 1967, the *Devonshire-Parker Fire* (Malibu Fire) burned 23,023b acres in the Santa Monica Mountains. But it didn't burn the camp.

Assigned to recuperate homebound for a month after the wildfire, I was miserable. That month contained: two weeks nursing a relentless headache with hot pads, antibiotics, and Tylenol, waiting for an HMO referral, and two weeks convalescing after surgery. After two weeks of headaches, I finally visited a doctor. Shining an otoscope up my nostrils, the ENT specialist diagnosed a sinus infection and polyp disease. He said I had "Polyps the size of grapes." While undergoing the polypectomy procedure, the surgeon discovered a "bucket of pus" that he drained.

At least my recovery from the polypectomy was far more enjoyable than the deviated septum procedure-a submucosal resection-seven years before at Doctors Hospital. I had a TV to occupy my time.

On December 1, I was back at work on a typical Saturday night. I was the DPO in charge waiting at the admin. office when the movie bus kids returned. Usually, the 17 top-ranked students plus one boy voted "most improved" by staff rejoined camp hyper from eating candy and the excitement of the monthly trip in street clothes to a first-run show thanks to a Hollywood producer that supported our program. But this night, rather than rushing back to the dorms, everyone quietly gathered around me. When I asked if anything was wrong, the pale face, John the "Ghost" Casper, dressed in fancy western duds, silver blond hair accentuating blue eyes, looked up, breaking silence, and said, "Push Broom, **Krazy Kamp** isn't such a **bad place**, after all."

Perplexed at the students' subdued manner and Casper's sincere tone, I said, "Casper, what movie did you see?"

"Cool Hand Luke!"

Made sense.

Perhaps due to the movie or maybe the cooler weather, attempted AWOLS subsided until Christmas weekend. Normally, among American stereotypical families, Christmas is a celebratory day of happiness and joy in exchanging gifts. However, for incarcerated children separated from family, Christmas is a time when their self-esteem plummets, anger turns inward, and loneliness and depression abound. Others lash out inciting reprisal as a form of self-blame for the predicament.

I was lucky enough to work on Christmas Eve, 1967. The phone rang at 0855 hours. "Wheeler," I answered, phone breaking my trance staring at the moonless sky through the admin's tinted glass panoramic window. Sounds of raucous youths echoed in the background.

"Jerry," Benny said, "I think Grant contracted a case of pre-release jitters."

I crossed the recreation field, scanning the surrounding landscape revealed in beams of powerful floodlights: smoldering thickets, burnt stumps, skeletal trees, and firebreaks covered with ashes. Since the fire, the wildlife hadn't come back. Some left for greener pastures while others were casualties of the recent conflagration: their hidden carcasses lay among the boulders. We tried to overlook the rotting meat but epically failed when the winds shifted.

When I entered C-Dorm at Benny's scheduled break, boys were sitting cross-legged on the floor, watching *Bonanza* on the Control Center's countertop mounted TV. Others were playing *ping pong,* and board games lying supine on their bunk daydreaming, reading paperbacks and comic books. Suddenly, Grant, a week from official discharge, decided to climb and stand in the middle of the pool table in his BVDs, screaming profanity, slapping the palm of his hand with a rolled-up towel, in a child version of Edward G. Robinson. His DPO counselor *Choo C*hoo gave staff the heads-up on Grant's pre-release anxiety uncommonly exhibited in regressive acting out. Behavior clinically attributed to youth's fear of returning to a previous unstable home environment marked by parental conflicts such as an authoritarian father, a submissive mother, intensive sibling rivalry, and school problems. Indeed, the counter-therapeutic milieu of *Krazy Kamp* was founded on *Summerhill's* philosophy of freedom without a license, democratic principles, and respect for individuality.

While Grant danced wildly on the pool table in his underwear, the dorm got caught up in the contagion. Boys commenced lobbying towels, socks, and T-shirts. Normally in these situations, the duty DPO gives the culprit some slack, time to calm down, hands him a write-up pad, instructing him to determine the punishment such as washing dorm windows, or police outside area. But Grant, joining the fray, wound up sentenced and threw a rolled-up towel at his mates exchanging fire, shouting, "Riot in C Dorm! Riot in C Dorm!"

Stern-faced, standing ramrod straight in front of the desk, arms at my side, dressed in wrinkled denim work shirt uncharacteristically draping over the belt of faded Levi's, I turned off the TV, then directed my attention to a small group of boisterous boys, blurting, "Riot in C Dorm! Riot in C Dorm!"

I projected my voice deeper than usual, I said, **"Quiet!"** scanning a dozen boys brandishing pool cues in the bunk area. The dorm grew silent. "I want everyone standing by his bunk, . . . **now!**"

Grant continued to ignore my instructions, dancing on top of a pool table for attention. Half the eyes focused on him. The other half was divided between the sword fighters and the Senior DPO giving the order to stand down.

In a raised voice, I said, "Grant, you have ten seconds to come down from there, put on your pants, and stand by your bunk until I make up my mind about where you are going to sleep tonight. It's either that or being escorted in your underwear to Intake Dorm."

"So whatchu gonna do, **Mister Wheeler**, if we riot?" Interrupted Charles Cardoza, an Italian kid, seething over a demotion from KP to Sweeper after he was caught smoking in the gym.

"Tell me something, Cardoza, are you a hermaphrodite?" His frowning bulldog face and silence conveying puzzlement, I said, "Then Cardoza, why are going around fucking yourself?" My words triggered loud laughter. My strong follow-up was, "Riot is not an option."

Not understanding his peers' reaction, Cardoza snapped, "Ya can't do nuttin."

No one moved. Even Grant, the one who started it all, remained silent, becoming a statue in shorts.

Waving a notepad, I said, "Know what this is, gentleman? A roster for the county bus to **YA**. If your name goes in here, you might as well pack up."

"You can't do that," Melvin Hooks barked, his black face reflecting a yellow film under the glare of the ceiling light in the control center.

I looked at the wall clock. "I can for the next nine hours and forty-six minutes. I am the Senior DPO-in-charge." A few feet started stirring.

"There's too many of us," blurted a hidden voice.

I spoke slowly. "You want the odds, gentlemen. You want odds. First, there is me." I smiled, looking around the dorm, then toward the admin. "Then there is **Popeye.** If I dare disturb his sack-time, you know what that's like. After him, there's the **Santa Monica Police**, a slew of **Sheriff Deputies,** the **State Highway Patrol**, the **National Guard**, and finally, the **Hundred and Second Airborne**. And gentlemen**, paratroopers** like to come in with **fixed bayonets."** I paused to give my charges time to reflect. "If that happens, let me assure you, your ass will no longer be in the hands of any Camp Kilpatrick official. Now, you all have had your little fun tonight. Anyone leaving this dorm without my permission better pack his personals 'cause he's not coming back."

Grunting and groaning, boys slowly started returning to their bunks in twos and threes. All had completed the Intake program and knew several things about Senior DPO Wheeler, a.k.a. *Pushbroom*: He never bluffed about discipline with them or other DPOs; never let a big kid assault someone smaller or a DPO manhandle a resident unless he was breaking up a mean fight. (The only occasion I resorted to physical force was when I cold-cocked a newbie brute that refused to stop waling on a kid half his size in the Intake Dorm's day room. During my two-year tenure, he was the only youth staff unanimously recommended transferring to YA.)

Reflections

Observing ideal adult interaction with male adolescent outliers fighting for self-identity, autonomy, and the space to allow mother nature to take her course, the DPO counselor Greg *Popeye* Benny, Rec Leader Keith Campbell *"KC"* Barton who performed DPO duties except for a caseload and completing court progress reports was a master.

When a new kid experienced the shock of being catapulted into an isolated, strange, mountainous desert devil's island called *"Krazy Kamp,"* first saw a resident "throwing a crazy": perched in a tree, straddling a boundary fence, being dragged kicking and screaming across a field, or scared of the dark and hungry hiding on the wrong side of the boundary fence, it didn't take long before the wasps in his stomach drove him within KC Barton's protective shadow, where he could feel the clouds of fear and loneliness lifted by KC's innate ability to connect to his frequency. Results often came without KC uttering a damn word but by his mere presence. On the other hand, when hounded by a gang of over-demanding clinging boys, KC's usual tolerant manner instantly changed to that of a Marine drill sergeant eliciting order.

Unlike KC, Benny rarely became flustered when dealing with a rash of kids throwing crazies or pulling at his sleeve demanding attention. *Popeye* Benny was also a master of the group situation, having a knack for keeping kids on track and drawing them in without them realizing it. During tirades, he took over where other DPOs left off. KC rarely reacted or got on one's case after he shot his bolt, whereas Benny followed my practice of waiting till a kid ran out of steam and then asked, "How do you feel about that?" And the kid would glare at him as if he were nuts or blind, wringing his hands, shaking his head, and remaining mute.

Then Benny or I would say, "Mad, pissed off, angry?" and we'd continue digging,

"Is this **camp** or **home** territory?"

If the former, such as a run-in with a teacher or DPO, or being harassed by another student, we would press the kid to describe his anger, go over the details of the incident and come up with alternatives to "throwing a crazy." We both harped on how hot-tempered youngsters must learn to "grow more skin," and take steps to "de-escalate" the situation. If the conflict was sparked by "home business," such as problems with parents missing visits, or brothers or sisters taking over one's space at home, we would encourage him to open up with his counselor or another DPO if the former was off duty.

Whether in group meetings or individually, KC and Benny quickly responded to lessons in anger management, de-escalation, "freedom not license," and "benevolent intervention," whereby another DPO intervenes during DPO/student conflict as other DPO retreats to defuse the confrontation. In the early stages of transforming the Intake program, a few DPOs argued that our kids need **more,** not **less,** structure, and self-governing reforms which violated common sense. The most effective reform advocates were ex-newbies hounding DPOs for more recreation resources such as ping pong tables and amenities available in Intake Dorm whence they came. My rejection of the notion of low expectations and risk avoidance for youth was attributable to being saved from a restrictive regimen of Catholic education in my final high school years and free rein in leisure time by guardians. From birth, I landed in the diverse, non-judgmental Carney clan's safety net with holes: Food & shelter, access to education, lessons in grief, rejection & rescue, mental illness, prejudice, racism, disabilities, family secrets never revealed, empathy, compassion, death, and ultimately-self-reliance. And perhaps it was the experiences of uprooted childhood, cathartic football head-ons, running away, white privilege dispensations granted youthful transgressions, OCS training, and academic and street mentors that helped me just as much.

I relished adventure, the "taking charge" of a rehabilitation program in relative shambles located in an isolated desert mountain setting occupied by a group of wild, tree-hanging, screaming, and rock-throwing child prisoners-of-war of a society that wanted them out of the way. Wayward boys were supervised by tuckered adults trying to referee. Liberal college-educated young men whose 62-hour shifts ended with zombie-like states. This suggested that they had reached their wit's end having surrendered to the forces of anarchy and their low expectations of the boys. But over time, two years of transforming a waystation into a deferential treatment model centered on A.S. Neill *Summerhill's* philosophy significantly raised both staff and resident expectations.

The ultimate proof for young boys and staff alike was demonstrated in the lack of panicking during the wildfire. During the transition, DPOs bonded more with kids and stepped outside their comfort zones such as Assistant Director Tom Gilder *Frankenstein* belatedly organizing a hiking club, *Bouncer* heading up arts and crafts, and *Tippy Toes* utilizing the Sony tape recorder during group sessions.

Wednesday Child

In 1968 California juvenile status offenders came under the purview of County Departments of Public Social Services (DPSS). Camp Kilpatrick's consultant clinical psychologist advised me that L.A. County DPSS was recruiting Child Welfare Supervisors for the Long Beach regional office. Given L.A. County Probation Department's rigid civil service criteria, the prospect of promotion to Assistant Director at Camp Kilpatrick was nil. Whereas DPSS rewarded MSW credentials and clinical experience. In addition, the substantial pay raise, closer work site, increased social justice activism, and family access with a baby on the way were incentives to transfer county departments close to home.

Children of poverty make up the bulk of dependency and delinquency cases. Child-welfare agencies alone placed over 300,000 children in 1967. If this number is combined with the number of delinquents-500,000-detained annually in overcrowded juvenile facilities, police lockups, and county jails, the total number of children at risk taken away from their families was approaching one million a year. The anguish of their lives is a repudiation of American values of human decency. [9]

The Long Beach DPSS Child Welfare unit was housed in the dingy basement of an ex-auto dealership on Long Beach Blvd. I oversaw five caseworkers assigned to 30-40 abused or neglected children. Our responsibility commenced within 24 hours of court-ordered placement. The minimum educational qualification of a child welfare worker was a college degree. My staff was composed of four white women: three under age 30, one over age 60– and a white male, in his mid-20s. Notwithstanding documenting abject poverty in-home visits in rural Ohio and South Central L.A., reviewing gruesome photographs of juvenile patricide, and "third world" disease depicted in Dr. Watson's medical books, by comparison, nothing compared to live images of malnutrition, bruises, burns, scars, and x-rays of broken bones of tortured infants and children taken at hospitals' ERs.

In 1963, California's Child Abuse and Neglect Reporting Act (Penal Code §§ 11164 - 11174.3) imposed child abuse reporting requirements on certain supervisors and human resources professionals. By law, teachers, social workers, and others were required to report child neglect and abuse to authorities like social services and law enforcement. In 1967 the U.S. Children Bureau reported: "One-third (202,333) of the children receiving welfare services were in foster homes, 10 percent (60,790) in institutions, and 7 percent (42,553) in adoptive homes."* Thus, a significant number of these children were removed from their homes by the juvenile court and child welfare agencies because their parents were too poor to provide care for them. Of the 452.7 million dollars spent on child welfare services, little or none went to families in the form of direct financial assistance. A child in trouble either made it on the welfare dole at home or was destined for placement. These thousands of placed children could be best described as America's orphans of the living.

Both dependent and delinquent children that come under the protection of the juvenile court have an equal chance of being taken away from home. Court decisions generally reflect the expectations of its delegate agencies. If the child-welfare worker or probation officer proposes for the child a one-sided "casework decision" which precludes providing more money to the client's needy family, there is a high probability the youngster will find himself in a foster home, group home, or institution. Yet the cost of placement or institutionalization for a single child far exceeds the amount of financial assistance allowed to him in his own home.

My mission was to review the staff's chronological narratives on unannounced home visits, treatment plans, and court progress reports. This involved listening to the caseworker's perception of the child's expressed words and body language, observing child-parent, guardian or foster parent interaction, and education needs., Child Welfare workers are selected for their people skills, and interest in helping children. Many were raised in middle-class families "safe homes;" never abused or neglected. In the mid-60s, Federal and State welfare laws gave considerable latitude and fiscal support to delegate agencies. Medical, dental, eye care and tutoring needs prescribed by the Child Welfare Supervisor were generally approved by the family court. I started monthly group meetings of our unit's foster care parents of adolescents with a psychiatric consultant.

Strict protocols like surprise visits and vigilant monitoring of abused children returned to biological parent(s) were a major part of our job description. In our experience, the worst cases were associated with the maternal mother's intoxicated boyfriend, stepfather, or unemployed biological father reacting to economic stress. Anyone, especially a male could wreak havoc on a child's life; even a laid-off aircraft engineer could lock kids in dark closets or impose hours of "wall time." In one instance, a stepfather forced a child to sit for hours in the ashes of an unlit fireplace. The most difficult cases involved conflicted children or toddlers who were covered with belts and burn scars while crying to return to the same biological father (their "daddy") who tortured them.

My initiation in child welfare supervision was changing my first infant's diaper at my desk under the watchful eye and instructions of a veteran child welfare worker. He was an infant found swaddled in a dishcloth and left at the entrance of the local hospital's ER. The legacy of "Baby Boy Doe" born without a birth record is terminal psychological rejection. Another "Baby Doe" in the orphan's purgatory: D.O.B. and name arbitrarily assigned by strangers. In winter, he wanders into the forest, cast away by icy limbs, and reaches for grapevines, dangling on frozen ground. In spring, no wildflowers, grazing deer, or flashing cardinals. No summer shade or glinting pristine pools, fingers bleeding from grasping thorn trees hiding father's secrets, and hives of swarming bees. Fall, no bed of leaves. Over the years, he backtracks in his nightmares and dreams, hacking through the rainforest searching for nests stolen by storms, trail markers of rocks, and bones, and lost tombstones of buried ancestors. Unlike a foundling searching for identity, I was given a surname on a birth certificate, a gift of the unbreakable human bond by diverse maternal kin, mentors, and strangers, in journeys crossing the Scioto River to heal a troubled land.

*1Eveline M. Burns, ed., Children's Allowance and Economic Welfare of Children (New York: Citizen Committee for Children of New York, 1968), p.3.

THE END

Made in the USA
Monee, IL
07 November 2022

17281808R00155